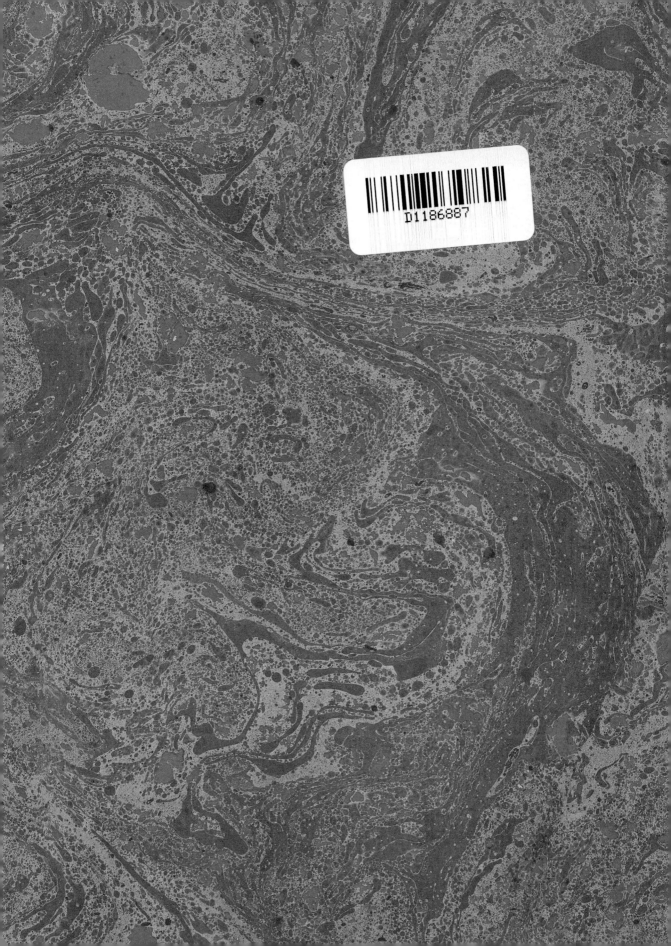

BUTTERFLIES
and MOTHS

BUTTERFLIES
and MOTHS

Engravings by
CHRISTIAN SEPP
and his son
JAN CHRISTIAAN SEPP

Text by
Dr. Stuart McNeill

MICHAEL JOSEPH
London

First published in Great Britain by Michael Joseph Ltd
52 Bedford Square, London WC1B 3ED 1978

© 1978 The Felix Gluck Press Ltd, Twickenham

Designed and produced by the Felix Gluck Press Limited

Photoset by Pierson LeVesley Ltd, Oxshott, Surrey
Printed in Switzerland by Hallwag AG, Bern

ISBN 0 7181 1688 7

Among the thousands of plates of Lepidoptera that have been produced over the last three hundred years, those of Christian and his son J.C.Sepp are outstanding not only for their technical accuracy, but also for their artistic merit. After Jan Christiaan's death, the Dutch Lepidoptera was continued by a number of other people, notably Brant who had a wonderful technique, but the standard set by the Sepps was never to be quite matched. I have, therefore, confined myself to the first seven volumes—those produced by Christian and J.C.Sepp themselves.

Separate plates illustrating, as colour forms of the same species, the complete life histories of closely related pairs of species which were not recognized as such until a hundred years or more later, show what exceptional observers and naturalists the Sepps indeed were.

Three main aims have directed my choice of plates. The first has been to relate the book to this country. The second to cover, as far as possible, some of the commoner species in all the major groups of the Macrolepidoptera, although I have been unable to do so with some families—e.g., the Lycaenidae (blues) and the Hesperidae (skippers) —of which the Sepps illustrated only a few species or none at all. My third aim has been to pick, whenever possible, the best available plates. This has led me at times to pass over a good plate of a very common species in favour of a superb plate of a slightly less common one.

The notes accompanying the plates give an outline of the species' life history, current status, habitat, range and distribution in this country. Their purpose is to make this book useful to the field naturalist and casual observer alike, in both town and country, and to bring these superb plates to the wide audience they deserve.

Stuart McNeill

CONTENTS

Foreword by Dr Stuart McNeill   9
Introduction by Ruud Rook   14

BUTTERFLIES

Swallowtail Butterfly   *Papilio machaon*   26
Large White Butterfly   *Pieris brassicae*   28
Small White Butterfly   *Pieris rapae*   30
Green-veined White Butterfly   *Pieris napi*   32
Orange Tip Butterfly   *Anthocharis cardamines*   34
Brimstone Butterfly   *Gonepteryx rhamni*   36
Comma Butterfly   *Polygonia c-album*   38
Small Tortoiseshell Butterfly   *Aglais urticae*   40
Peacock Butterfly   *Inachis io*   42
Painted Lady Butterfly   *Vanessa cardui*   44
Painted Lady Butterfly (contd)   *Vanessa cardui*   46
Red Admiral Butterfly   *Vanessa atalanta*   48
Dark Green Fritillary Butterfly *Mesoacidalia aglaja*   50
Marsh Fritillary Butterfly   *Euphydryas aurinia*   52
Glanville Fritillary Butterfly   *Melitaea cinxia*   54
Speckled Wood Butterfly   *Pararge aegeria*   56
Wall Brown Butterfly   *Lasiommata megera*   58
Wall Brown Butterfly (contd)   *Lasiommata megera*   60
Meadow Brown Butterfly   *Maniola jurtina*   62
Gatekeeper Butterfly   *Pyronia tithonus*   64
Ringlet Butterfly   *Aphantopus hyperanthus*   66
Small Heath Butterfly   *Coenonympha pamphilus*   68
Purple Hairstreak Butterfly   *Quercusia quercus*   70
Silver-studded Blue Butterfly   *Plebejus argus*   72

HAWKMOTHS

Lime Hawkmoth	*Mimas tiliae*	74
Poplar Hawkmoth	*Laothoë populi*	76
Eyed Hawkmoth	*Smerinthus ocellata*	78
Death's Head Hawkmoth	*Acherontia atropos*	80
Death's Head Hawkmoth (contd)	*Acherontia atropos*	82
Convolvulus Hawkmoth	*Herse convolvuli*	84
Convolvulus Hawkmoth (contd)	*Herse convolvuli*	86
Privet Hawkmoth	*Sphinx ligustri*	88
Privet Hawkmoth (contd)	*Sphinx ligustri*	90
Pine Hawkmoth	*Hyloicus pinastri*	92
Elephant Hawkmoth	*Deilephila elepnor*	94
Elephant Hawkmoth (contd)	*Deilephilia elepnor*	96
Hummingbird Hawkmoth	*Macroglossum stellatarum*	98

MOTHS

Puss Moth	*Cerura vinula*	100
Lobster Moth	*Stauropus fagi*	102
Lobster Moth (contd)	*Stauropus fagi*	104
Swallow Prominent Moth	*Pheosia tremula*	106
Pebble Prominent Moth	*Notodonta ziczac*	108
Buff Tip Moth	*Phalera bucephala*	110
Peach Blossom Moth	*Thyatira batis*	112
Vapourer Moth	*Orgyia antiqua*	114
December Moth	*Poecilocampa populi*	116
Oak Eggar Moth	*Lasiocampa quercus*	118
Oak Eggar Moth (contd)	*Lasiocampa quercus*	120

MOTHS (contd)

Fox Moth	*Macrothylacia rubi*	122
Fox Moth (contd)	*Macrothylacia rubi*	124
Fox Moth (contd)	*Macrothylacia rubi*	126
Drinker Moth	*Philudoria potatoria*	128
Emperor Moth	*Saturnia pavonia*	130
Emperor Moth (contd)	*Saturnia pavonia*	132
Pebble Hooktip Moth	*Drepana falcataria*	134
Green Silver-lines Moth	*Bena prasinana*	136
Buff Ermine Moth	*Spilosoma lutea*	138
Ruby Tiger Moth	*Phragmatobia fuliginosa*	140
Garden Tiger Moth	*Arctia caja*	142
Cream-spot Tiger Moth	*Arctia villica*	144
Cinnabar Moth	*Tyria jacobaeae*	146
Nut-tree Tussock Moth	*Calocasia coryli*	148
Grey Dagger Moth	*Acronicta psi*	150
Powdered Wainscot Moth	*Simyra albovenosa*	152
Flame Shoulder Moth	*Ochropleura plecta*	154
Large Yellow Underwing Moth	*Noctua pronuba*	156
Large Yellow Underwing Moth (contd)	*Noctua pronuba*	158
Dot Moth	*Melanchra persicariae*	160
Bright-line Brown-eye Moth	*Diataraxia oleracea*	162
Campion Moth	*Hadena cucubali*	164
Green Brindled Crescent Moth	*Allophyes oxyacanthae*	166
Merveille du Jour Moth	*Griposia aprilina*	168
Angle-shades Moth	*Phlogophora meticulosa*	170
Copper Underwing Moth	*Amphipyra pyramidea*	172
Mouse Moth	*Amphipyra tragopogonis*	174
Hebrew Character Moth	*Orthosia gothica*	176

Common Quaker Moth	*Orthosia stabilis*	178
Dunbar Moth	*Cosmia trapezina*	180
Red Swordgrass Moth	*Xylena vetusta*	182
Beautiful Yellow Underwing Moth	*Anartia myrtilli*	184
Bordered Sallow Moth	*Pyrrhia umbra*	186
Herald Moth	*Scoliopteryx libatrix*	188
Burnished Brass Moth	*Plusia chrysitis*	190
Silver Y Moth	*Plusia gamma*	190
Mother Shipton Moth	*Euclidimera mi*	192
Red Underwing Moth	*Catocala nupta*	194
Snout Moth	*Hypena proboscidalis*	196
Large Emerald Moth	*Geometra papilionaria*	198
Mallow Moth	*Larentia clavaria*	200
Streak Moth	*Chesias legellata*	202
Winter Moth	*Operophthera brumata*	204
Phoenix Moth	*Lygris prunata*	206
Spinach Moth	*Eulithis mellinata*	208
Common Marbled Carpet Moth	*Chloroclysta truncata*	210
Silver Ground Carpet Moth	*Xanthorhoë montanata*	212
Magpie Moth	*Abraxas grossulariata*	214
Large Thorn Moth	*Ennomos autumnaria*	216
Swallowtailed Moth	*Ourapteryx sambucaria*	218
Brimstone Moth	*Opisthograptis luteolata*	220
Brindled Beauty Moth	*Biston hirtarius*	222
Oak Beauty Moth	*Biston stratarius*	224
Peppered Moth	*Biston betularius*	226
Peppered Moth (contd)	*Biston betularius*	228
Bordered White Moth	*Bupalus piniaria*	230
Six-spot Burnet Moth	*Zygaena filipendulae*	232
Hornet Clearwing Moth	*Sesia apiformis*	234

The abbreviation appearing after the scientific name in the headings is a standard contraction of the name of the person who first described the species under the given specific name; e.g. L. is for Linnaeus.

INTRODUCTION

A great interest in plants and animals flourished in Holland in the second half of the seventeenth century. This was the period of Jan Swammerdam (1637-1680), the man who discovered the red corpuscle, and Antoni van Leeuwenhoek (1632-1723), a pioneer in microscopy, and this was also the time of the great natural history collections and 'rarity cabinets'. Jan Swammerdam's father, a wealthy chemist in Amsterdam, possessed a rarities cabinet he believed to have a value of sixty thousand guilders; and Jan's colleague Seba takes up four thick folio volumes to describe his own collection.

By the seventeenth century the Amsterdam book trade was already showing an interest in natural history. In 1660 Jonston's account of the nature of the quadruped animals, birds, etc., was published in that city.

In the first half of the eighteenth century Maria Sibylla Merian (1647-1717) obtained financial backing for her voyage to Surinam, a Dutch colony at that time. The drawings of caterpillars and pupae of butterflies which she made there are now world famous. During her time in Surinam (1698-1701) she also did drawings of food plants of butterflies for her major work on the reproduction and miraculous metamorphosis of Surinam insects, *Metamorphosis insectorum Surinamsum*, which contained 71 plates coloured by hand. (In the eighteenth century this was the only way of reproducing colour plates in books and the Dutch artists excelled at this work.) The first Dutch edition of this wonderful book was published in 1705; within a short time it had gone into three impressions as well as German and French editions. This, together with her second book on European insects, *Europäische Insecten*, with 184 engravings on 47 hand-coloured plates, published by her daughter Dorothea Maria Henrietta, are today considered to be among the finest insect books of the eighteenth century.

The now famous *Mémoire pour servir à l'histoire des insectes* by René Antoine Ferchauld de Réamur (1683-1757), published 1737-1748 by Pierre Mortier, is evidence of Amsterdam's importance as a city of international publishing. This work was Réamur's most important contribution to science. It describes the appearance, habits and habitat of all insects known at the time, with the exception of beetles, and is a marvel of patient and accurate observation. Particularly noteworthy are his 'mémoires' on bees and the experiments that enabled him to prove the truth of Peysonel's hypothesis that corals are animals and not plants.

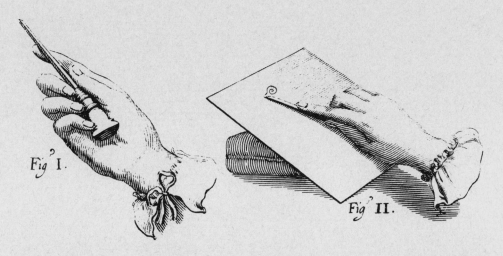

Above: illustrations from Bosse's seventeenth-century book on engraving techniques. Fig I *an engraver's burin;* Fig II *the use of a leather cushion under the copper plate.*

It is not, then, altogether surprising that Amsterdam, as magical a place then as it is now and one of the eighteenth century's richest cities, should have attracted many artists and enterprising men who made their living by illustrating books on natural history. Already by 1680, when the famous copper engraver Romeyn de Hooghe left Amsterdam to settle in Haarlem, there were 110 other engravers active in Amsterdam.

Books first began to be illustrated with copper engravings in the seventeenth century, but although the technique of copper engraving had been discovered as early as the fifteenth century, wood engraving remained the main method of illustrating books until the beginning of the eighteenth century. The popularity of scientific works and the increased efficiency of the printing presses were the main factors contributing to the supersedure of wood engraving by etching and copper engraving—the quality of wood engraving depended so much on the grain of the wood, while copperplates were much more regular and flexible.

The technique used in the making of copperplates was as follows. A plate of copper approximately 2.5 mm thick was polished as smooth as possible. The original drawing, reversed, was then transferred to the plate. A burin (as illustrated above) was used to follow the lines drawn on the plate. Its fine chisel-like point pushed out of the soft copper a thin thread of the metal. Because this tool could only move in one direction, the copperplate was set on a round leather cushion on which it could be

15

rotated to enable the burin to follow curved lines. The etching process increased the possibilities of including finer lines, adding to the delicacy of the illustrations.

With regard to hand colouring, woodcuts had constrained the colourist to colour within an existing framework of lines, almost like today's colouring books. The fine lines of the copper engravings gave the colourist a chance to add depth to his work and bring the illustrations to life, as is exemplified by Sepp's work.

16

An illustration from Diderot's 'Encyclopédie' shows an eighteenth-century engraver's workshop. The copper plate is first warmed (Fig 1) before being varnished. The varnished surface is then smoked (Fig 1 bis). Figs 3, 4 and 5 show different ways of handling the acid that etches the design into the metal; in Fig 6 an engraver uses a burin to cut directly into the copper plate. The plate may easily become warped during the etching process—Fig 7 shows a plate being flattened.

The Sepp family and their books

Christian Sepp was born about 1700 in Goslar, Germany, where his father was senior master of the grammar school. Intended for a career as a merchant, he settled in Hamburg as a dealer and commission-agent, although in those early days he already had a great love of the arts and sciences, especially physics. While in Hamburg he prepared and constructed a number of wooden and copper scientific instruments, which he was able to make use of in his entomological research. Evidence of another of his talents is given by one of Sepp's contemporaries, who was so impressed by the quality of his handwriting at the time that he thought Sepp must be capable of making fine drawings.

We cannot be sure how long Christian Sepp lived in Hamburg, but we do know that he was married in Germany and that he settled in Amsterdam before 1739 when his first son Christiaan was born.

In Amsterdam Christian Sepp took up the etching needle and burin and established a name for himself by drawing land and sea maps, though he was to win his greatest renown by his knowledge of Dutch insects and the excellent drawings which he engraved for his masterpiece, *Beschouwingen der wonderen Gods in de minst geachte schepselen of Nederlandsche insecten naar hunne aanmerkelijke huishouding, verwonderlijke gedaanteversisseling en andere wetenswaardige bijzonderheden beschreven* (observations of the wonders of God, the least regarded of Dutch insects, including descriptions of their remarkable habitat, their surprising metamorphosis and other details worth knowing)—abbreviated to *De Nederlandsche Insecten* (Dutch insects).

This was the published in the form of what we would call a 'part work' and sold on a subscription basis. The first series of parts was published in 1762. Many of the plates in this first series bear Christian Sepp's special mark: 'C. Sepp ad viv. et sculpsit'—drawn from nature and engraved. The 1762 series consisted of five separate parts containing a varied number of descriptions of butterflies and moths, totalling 50 in all. A list of contents was given with the last part in that year, thus enabling the collector to sort the articles into the correct sequence and to have them bound by a bookbinder. It was customary in those times for the individual to have his books personally bound, and some of the most beautiful bookbindings were done in that period.

In 1762 the price for the set of five part works making up the first volume of Sepp's book was 47 Dutch guilders, which would be about

£120 at today's purchasing value. It was a fair price however, considering the precision and craftsmanship demanded in the hand colouring of these plates.

In his introduction Christian Sepp writes that in the future he intends also to include beetles, grasshoppers, wasps, flies, etc., but this promise he never realized. *De Nederlandsche Insecten* was to remain a book about butterflies and moths. In the same introduction Sepp states that his work is in no way an imitation of August Johann Rösel von Rosenhof's *Insecten-Belusstigung*, a well-known contemporary work on insects, containing finely hand-coloured plates. Sepp does admit a certain resemblance to that work, but immediately asserts that the similarity is only one of size and classification. On the other hand, he often refers to this work in his *Nederlandsche Insecten*, as also to Réamur's *Mémoire pour servir à l'histoire des insectes*.

The publishing 'privilege'—an early form of copyright granted for fifteen years by the States of Holland—mentions 'Christiaan Sepp en Zoon, Auteurs en Uitgevers van seker Werk' (Christian Sepp and son, authors and publishers of certain work). It is thought that the unsigned plates are by Christian's son Jan Christiaan, who continued to work alone after his father's death in 1775. But many plates and descriptions must be by the father himself, about whom Mr Alfred Sich wrote in the *Entomologist's Record* of January 1901, 'Sepp was an exquisite artist and an accurate observer . . . He had the touch of life in his brush.'

A few examples of the original text that accompanies the plates show how they were based on personal observation and not on other sources. In one, Sepp explains how he obtained eggs by catching a female that had mated in freedom. He got 150 eggs by this method, but was less successful when he tried to get them by putting males and females together:

> '*And now I put males and females together in the hope that they would mate and deposit eggs, but alas, I do not know whether they have done the former, for they died without having done the latter.*'

On another occasion he had better luck:

> '*Never did I have the luck to find the eggs of this moth, but by putting some female and male moths together in a box, I managed to get possession of some.*'

There seems to have been at that time a great deal of interest in butterflies and moths and their development, for Sepp clearly had a circle of like-minded friends who helped him in his research:

> '*I had not known what the eggs of the insect looked like, but finally those eggs, which I had wanted so long, were delivered to me through the kindness of the most honourable Mr van Bekestein Haket, so that I was able to complete the story of this insect.*'

Finally some observations on caterpillars:

> '*The colour of the eggs turned darker after having been in my house a few days. Shortly thereafter the caterpillars hatched, greyish in colour, with many fine hairs.*'
> '*While in the province of Overijssel, I found two fully grown caterpillars on the reed-like grass. They will have their metamorphosis next year and I will be able to learn their complete history.*'
> '*On August 11th the caterpillars had their first moult, so their black heads turned green. On August 18th they had their second moult, on August 27th their third moult, on September 4th their fourth and on the 15th their last one. Every time they ate their moulted skin.*'

In other parts of the text we find meticulous notes on the metamorphosis of his caterpillars. Sepp must have spent days and nights observing the caterpillars, drawing them and, finally, engraving them in copper.

From an early age Jan Christiaan Sepp showed a great interest in his father's studies and his craft. In the text by Christian which accompanies the plates there are many descriptions of how he and his son together found caterpillars, etc., and soon Jan Christiaan too was active with microscope and brush helping his father, drawing and describing butterflies, eggs and pupae. He also constructed with his own hands his own 'rarities cabinet' which, though small, was very much admired in his day.

When the Amsterdam authorities wanted a publisher and bookseller for *De Nederlandsche Insecten* it was Jan Christiaan Sepp who undertook the task. It is amusing to see the father doing a bit of propaganda on behalf of his son's book trade. In footnotes in *De Nederlandsche Insecten*, referring to some foreign titles, Christian Sepp adds that they are available in his son's shop at the same price as in their country of origin. One of these titles is Dr Schaffer's *Elementa Entomologica* or *Einleitung in die Insecten Kenntnis*, with 135 colour plates—again each one individually coloured by hand for every single copy of the book.

One of the first titles to appear on the publisher's list of Jan Christiaan Sepp was Casper Philips Jacobsz's work on perspective and optics, *Over de Perspectief of Doorzichtskunde*. Author and publisher must have had a very good relationship because on plate 42 is depicted, slightly hidden behind a winding staircase, Sepp's cabinet, which can today be admired in the Natural History Museum of Leiden.

The next major enterprise Jan Christiaan Sepp embarked upon was the publication of *De Nederlandsche Vogelen* (Dutch birds) (1770-1829). He was able to undertake this second major project thanks to the sales success of his butterfly book, which enabled him to engage master craftsmen who hand-coloured his engravings with amazing accuracy.

In 1768 Jan Christiaan married Sara Focking, the daughter of an Amsterdam paper seller. Their first two children died when young and in 1773 Sara herself died after giving birth to their only surviving son, Christian. In the same year Jan Christiaan married Wichertje Wichers Kruys of Giethoorn in the province of Overijssel, a province Jan Christiaan knew well from his butterfly-collecting expeditions—perhaps he and his wife met on one of these. Their marriage, according to contemporary reports a very happy one, was blessed with eleven children.

Shortly after Jan Christiaan's second marriage his father died. He was now the publisher, author, editor and engraver of *De Nederlandsche Insecten*. The Sepp family accepted the task of continuing this major work as a debt of honour: from Jan Christiaan it was to pass to his son Jan and from Jan in turn to his son Cornelis, so that in all four generations devoted themselves to it.

At the time of his father's death, Jan Christiaan Sepp was publishing Pieter Cramer's *Uitlandsche Kapellen* (Foreign butterflies) and the Dutch translation of Edwards and Catesby's bird book, both of which were great commercial successes. Further proof of Jan Christiaan's love for entomology came when he published Casper Stoll's book on cicadas and bugs, bush and migratory crickets. This was followed by Stoll's famous and highly prized work, *Natuurlijke en naar het leven nauwkeurig gekleurde afbeeldingen en beschrijvingen der spoken, wandelende bladen, springhanen, enz.* (Pictures that are true to nature, drawn from life and accurately coloured, with descriptions, of grasshoppers, leaf insects, saltants, etc.), in two volumes with 70 fine hand-coloured plates.

Another achievement of Jan Christiaan Sepp was 'Felix Merites' (Happiness by merit), a society for artists and scientists which he and his friend Willem Writs founded and which continued to exist until 1888. And in 1781 Jan Christiaan was elected pastor of a Mennonite parish—a fact which again underlines the significant social role he played.

One final great enterprise undertaken by J.C. Sepp lay in the field of botany. His firm had already published the Dutch translation of a work on medical herbs, but now he embarked on the *Flora Batava of Afbeelding en beschrijving van Nederlandsche gewassen* (Flora Batava, or pictures and descriptions of Dutch plants) (1800-1934), a work that ultimately went to 28 volumes, with 2,240 plates, of which 1,920 were finely coloured by hand—the later ones, which were not published by Sepp, were in chromolithography. This work is to date the only fully illustrated flora of the Netherlands; the text is in Dutch and French. Jan Christiaan Sepp is probably responsible for the first eight volumes, which are the most finely illustrated. The Utrecht botanist, Jan Kops (1765-1849), was the first author; he was succeeded by several others.

Jan Christiaan Sepp died in 1811 at the age of seventy-two. He had been a great entomologist, an even greater artist, a highly successful publisher and bookseller and a man who played an active part in the social life of his day. In his *Suites à Buffon*, Guenée describes J.C. Sepp's butterfly pictures as unrivalled masterpieces.

In terms of typography alone *De Nederlandsche Insecten* is a work of great merit. In his *Anderhalve eeuw boektypografie, 1815-1965* (One and a half centuries of book typography), Professor Dr G.W. Ovink writes that Sepp's *Nederlandsche Insecten* is significant for the fact that the publisher continued to use the eighteenth-century typeface to match the style of the engravings until there was no longer a printer in the country who stocked these typefaces. This contributed greatly to the continuity and style of this work.

When Jan Christiaan Sepp died, his son Jan, born 1778, continued the Firma J.C. Sepp & Zoon. Jan was, like his father and grandfather, an entomologist and the publisher, editor and author of *De Nederlandsche Insecten*, but although he was a skilful artist he was not able to prepare all the engravings himself, perhaps because the firm's other publications took up too much of his time. Several amateur and professional artists made drawings for the work, the most talented of these contributors being Quirijn Maurits Rudolph Ver-Huell (1787-1860), who did the plates of the Mouse Moth (p 175), the Streak (p 71) and the Hornet Clearwing (p 235). Ver-Huell was a naval officer who enjoyed a certain fame for the fine maps he drew. In 1810 he had been sent on a spying mission to make correct maps of the river Eider and the Holstein Channel in Germany. He was a talented draughtsman, aquarellist, painter and entomologist. His *Handboek voor liefhebbers en verzamelaars van vlinders* (Handbook for amateur and professional butterfly collectors) was published in 1842—not by Sepp. He made hundreds of drawings for a great variety of books on natural history, among which *Flora Brasiliensis*, edited by Professor Martius, deserves a special mention.

A copperplate printer's workshop, from Diderot's 'Encyclopédie'. The wet prints hang on the wall to dry.

To return to the publisher, Jan Sepp: he completed the publication of Nozeman and Houttuyn's *De Nederlandsche Vogelen*, which contains within five volumes pictures and life histories of 250 bird species. Among his later productions must be mentioned *Afbeeldingen en beschrijvingen van Surinaamsche Vlinders* (Pictures and descriptions of Surinam butterflies) and *Bloemkundig Woordenboek* (Dictionary of flowers) generally known as 'Sepp's Bloemkundig Woordenboek' because although Sepp did not write it, the actual author was anonymous. This latter book was very popular in its day.

Jan's successor was his second son, Cornelis, born 1810. Cornelis Sepp was no entomologist himself, and he engaged Mr S.C. Snellen van Vollenhoven under whose editorship a second series of *De Nederlandsche Insecten* (not used in this book) was started.

23

In 1868, with the death of Cornelis Sepp, the Firma J.C. Sepp & Zoon publishing house ceased to exist—Cornelis's only son became a Mennonite minister. The butterfly book started by Cornelis Sepp's great grandfather and the *Flora Batava* started by his grandfather were taken over by two separate publishers.

The new publisher of *De Nederlandsche Insecten*, Martinus Nijhoff of The Hague, actually started a third series edited by M. Brants and illustrated with lithographs. The fact that this never attained the purity and beauty of the Sepp series was due no doubt to the newness of lithography as a technique, attractive because it was so much cheaper than the old engraving process with its labour intensive hand-colouring. It is quite amazing how accurate the hand-colouring of the early engravings had been, and how the masters of that craft had managed to reproduce the same colours with almost no variation from one book to the next.

In Cornelis Sepp's estate were a great number of drawings on entomological subjects by his great grandfather and grandfather, and a few by his father and others. These were bequeathed by his widow to the Dutch Entomological Society.

R.C. Kruseman, writing of Cornelis Sepp in *Bouwstoffen voor een Geschiedenis van den Nederlandschen Boekhandel* (Foundations for a history of the Dutch book trade), said, 'He was the last to carry the name of a famous family in the book trade. . . . For over a century this publishing house maintained its fame. . . . What Dutch publisher can rival the Firma J.C. Sepp & Zoon for bold initiative? Who has had such a wealth of talent to offer?' Firma J.C. Sepp & Zoon began in eighteenth-century Stadtholder's Holland, survived the revolutionary and Napoleonic years and the restoration, and continued well into the nineteenth century. Today we may well ask ourselves, how many publishers have managed to maintain such consistently high quality over such a long and chaotic period?

Ruud Rook

Right: an original title page from the first series of the Sepp part works on Dutch insects.

DE

WONDEREN

GODS

in de

minst geachtste

SCHEPSELEN.

THE SWALLOWTAIL BUTTERFLY

PAPILIONIDAE *Papilio machaon* L.

This handsome species, the largest of our native butterflies, is now confined to the fens of East Anglia, a small part of its former habitat. In other parts of its wide range abroad, this butterfly may have between one and three broods a year. However, in this country, there is only one main brood which is on the wing in late May and June. In particularly favourable years there is often a partial second brood on the wing in August and September.

The eggs (*Fig 1*) are laid on milk parsley (*Peucedanum palustre*) or sometimes wild carrot (*Daucus carrota*) and hatch after about ten to twelve days. The young caterpillar (*Fig 2*) is at first black with a small white patch, resembling a bird-dropping, but as it grows it becomes more highly coloured and conspicuous (*Fig 3*). By the time it has moulted three times it has assumed the colour of the fully grown caterpillar (*Fig 4*) and is a very conspicuous object on the food plant. At this stage it has developed a unique defence organ, the osmaterium (*Fig 8*) which is concealed in a pouch behind the head. This organ is bright orange and, when the caterpillar is disturbed, shoots in and out rapidly like the tongue of a snake.

The larva becomes fully grown in about six or seven weeks and moults to the pupal stage. This is often to be found, head uppermost, attached to reeds close to the larval food plant, held by a girdle of silk around the middle of the body. These pupae normally overwinter, but in a warm season may emerge in August or September.

Abroad this butterfly has a very wide range and is found in a much wider range of habitats than it is in this country. It is a well-known wanderer, and examples of the continental race sometimes turn up in southern counties.

1 Egg 2 Young larva 3 Larva 4 Larva
5 Pupa 6 Adult butterfly
7 Adult butterfly 8 Osmaterium
9 Cremaster (pupal hook-like attachment)
Plant: Wild carrot (Daucus carrota)

Fig . 6.

Fig . 8.

Fig . 4.

Fig . 2.

Fig . 1.

Fig . 7.

Fig . 9.

Fig . 3.

Fig . 5.

THE LARGE OR CABBAGE WHITE BUTTERFLY

PIERIDAE *Pieris brassicae* L.

This common butterfly has two broods per year and may sometimes have three. Adults are found in spring (April and May) and again in summer (July and August). The adult butterflies of these two broods differ slightly in colouring, the spring brood generally having a much less intense black colouring on the wings. The wing tips in particular have a tendency to become distinctly grey in colour.

Males and females differ in colour pattern, the male (*Fig 7*) lacking the black spots on the upper surface of the fore-wings and the black dash on the hind-wings of the female (*Fig 6*). Both sexes have two black spots on the under-surface of the fore-wings (*Fig 5*), although the lower spot is usually much weaker in the male. The illustration is of a female.

The eggs (*Fig 1* natural size, *Fig 2* enlarged) are laid in batches on a variety of species of Cruciferae, especially the genus *Brassica* which includes our garden cabbages and cauliflowers, and also on the garden nasturtium (*Trapaeolum*). The eggs hatch in about a week and the caterpillars feed up rapidly and spin up in sheltered places, especially fences and walls, near the food plant. The chrysalis is shown in *Fig 4*. In Britain only fairly small numbers of the pupae of the second generation succeed in overwintering and the population is maintained by immigration of adult butterflies from the continent.

The caterpillars of this species are often attacked by a range of insect parasitoids, viruses and bacteria. The commonest parasitoid is a small braconid wasp, *Apanteles glomeratus*. This insect lays its eggs in the young caterpillers of the Cabbage White. The eggs hatch and the parasite larvae feed up inside the body of the host which continues to grow apparently normally. However, when the time comes for it to turn into a chrysalis the parasite larvae, now fully grown, emerge from the body of the host, killing it, and spin their own distinctive yellow cocoons on the outside of the dead caterpillar. A single butterfly caterpillar may contain up to 150 parasite larvae.

1 Eggs, natural size
2 Egg enlarged to show detail 3 Larva
4 Pupa 5 Underside of female butterfly
6 Female butterfly 7 Male butterfly
Plant:Mustard (Sinapis sp)

Fig. 7.

Fig. 6.

Fig. 1.

Fig. 3.

Fig. 2.

Fig. 4.

Fig. 5.

C. Sepp ad viv. del. et sculpsit

THE SMALL WHITE BUTTERFLY

PIERIDAE *Pieris rapae* L.

This species is one of our commonest butterflies and has two broods each year. The first brood emerges from the overwintered pupae in May and June. These insects are usually more heavily marked than those of the second brood, which are on the wing in August and September. As in the Large White, male and female differ in the extent of the dark markings on the wings. The females (*Fig 8*) have more spots on the hind area of the fore-wings than do the males (*Fig 7*).

The eggs (*Figs 1, 2*) are laid on much the same range of host plants as those of the Large White, and these two species are often serious pests of field and garden Brassica crops, especially cauliflowers where the larvae get into the curds and cause extensive fouling. The pupae (*Fig 5*) are often found on wooden fences, shed walls etc. close to the feeding sites of the larvae. The adult butterflies of the summer generation emerge from the pupae after some two to three weeks, whereas the pupae from this generation of adults, formed in the autumn, overwinter in this stage. This butterfly is also often attacked by *Apanteles glomeratus*.

The Small White is found throughout Europe, North Africa and Asia to Japan, and has been introduced, and become a pest, in North America and Australia.

1 Eggs, natural size, laid on leaf
2 Egg enlarged to show detail
3 Larva 4 Larva 5 Pupa
6 Female butterfly 7 Male butterfly
8 Female butterfly
Plant: Nasturtium (Tropaeolum sp)

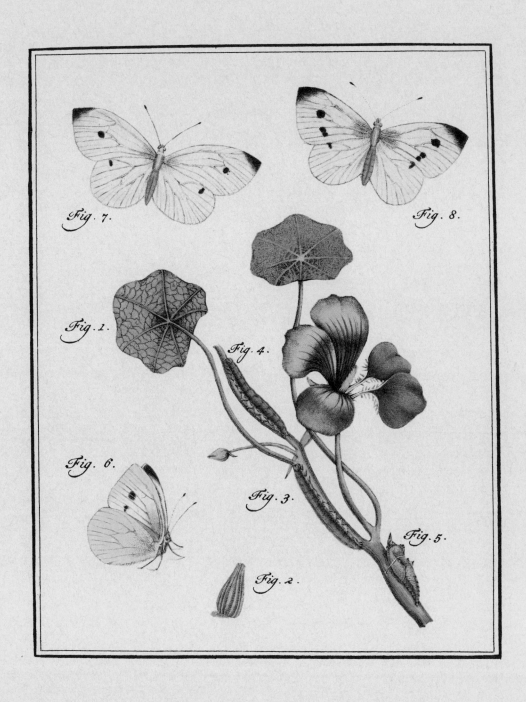

Fig. 7.

Fig. 8.

Fig. 1.

Fig. 4.

Fig. 6.

Fig. 3.

Fig. 5.

Fig. 2.

THE GREEN-VEINED WHITE BUTTERFLY

Pieridae *Pieris napi* L.

This is the third of our common species of white butterfly and, as with other members of this group, its life history is similar to that of the Large White. On the wing this species is often confused with the Small White, but a closer examination soon reveals the difference. The wing veins of the upper surface are powdered with black scales (*Fig 6*) and the lower surface of the wings has a distinct greenish banding along the veins, particularly of the hind-wings (*Fig 5*).

The eggs (*Figs 1, 2*) are laid on a number of cruciferous plant species such as white mustard (*Sinapis alba*) and charlock (*S. arvensis*) which seem to be preferred to the cultivated Brassicas. This is probably due to the later instars' preference for the plants' seed pods rather than their leaves. The larvae (*Fig 3*) feed up in six or seven weeks and spin up (*Fig 4*) nearby. Although a species associated with woodland edges rather than man-made habitats, it is often common in gardens.

This butterfly is rather more variable in colour than the preceding ones, and brownish and yellowish examples are much more common, the latter being particularly prevalent in Ireland. The two broods (May-June and July-August) differ in the intensity of the black markings and also between male and female as in the preceding species.

This species is found throughout Europe, North Africa and west and central Asia as well as in North America.

1 Eggs, natural size, laid on leaf
2 Egg enlarged to show detail 3 Larva
4 Pupa 5 Adult butterfly
6 Female butterfly
Plant:Charlock (Sinapis arvensis)

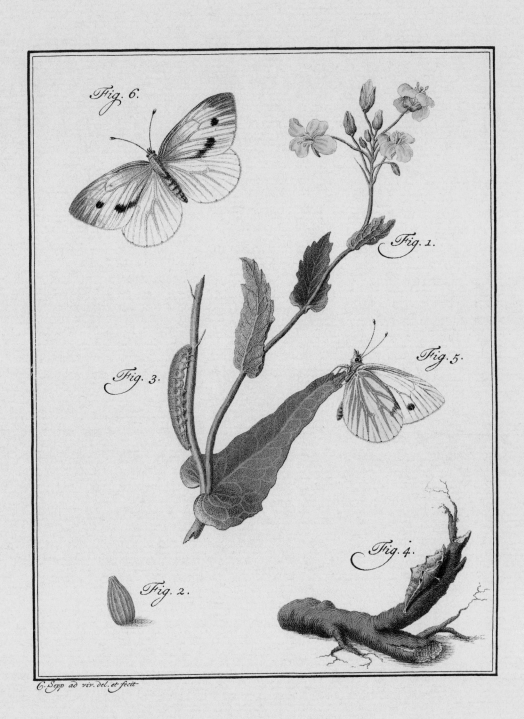

Fig. 6.

Fig. 1.

Fig. 3.

Fig. 5.

Fig. 4.

Fig. 2.

C. Sepp ad viv. del. et fecit.

THE ORANGE TIP BUTTERFLY

PIERIDAE *Anthocharis cardamines* L.

In this butterfly the sexes are even more distinct than in the preceding group of pierids. The bright orange tips to the wings which give the species its English name are found only in the male (*Figs 10, 11*), the female (*Fig 12*) being much more like a Green-veined White, although the markings are a mottling of the wings rather than along the veins. This species however, unlike the others, is only single-brooded with the adults on the wing from late April until June.

The eggs, which are similar to those of the Small White, are laid on cuckooflower (*Cardamine pratensis*) and a number of other similar plants. The attractive larvae (*Figs 1, 2, 3, 4, 5*) feed on the leaves, flowers and fruit of the plant; they are particularly difficult to find on the latter which they closely resemble in colour and form. The larvae are usually fully fed by August and spin up on a stem nearby. The pupa (*Figs 6, 7, 8*) is of an odd shape, being very elongated and angular, green at first (*Figs 7, 8*) but turning brown later. This stage overwinters to emerge the following spring. A small proportion of the pupae are known to emerge two years later, an adaptation common in many Lepidoptera to ensure survival in bad years.

This butterfly, although widely distributed, is more local than the previous species and can be seen along grassy rides and woodland edges as well as meadows where its food plant grows. It is found throughout much of temperate Europe and Asia.

1-5 Larvae 6-8 Pupae
9 Female butterfly 10 Male butterfly
11 Male butterfly 12 Female butterfly
Plant:Garden stocks (Mattiola sp)

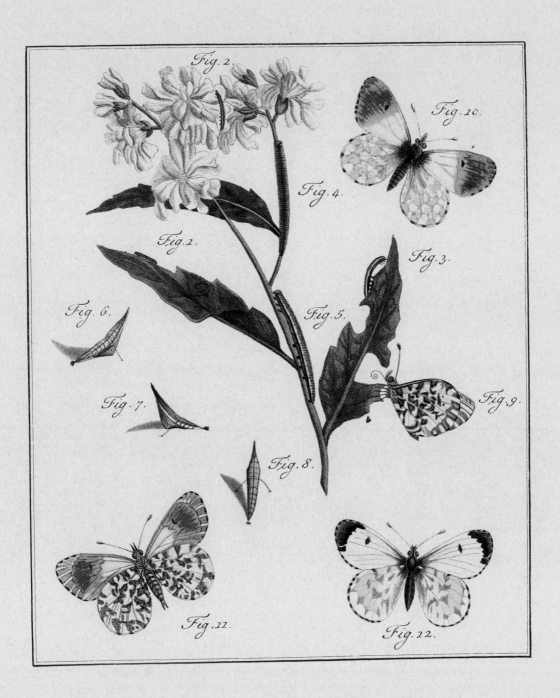

Fig. 2.

Fig. 10.

Fig. 4.

Fig. 1.

Fig. 3.

Fig. 6.

Fig. 5.

Fig. 9.

Fig. 7.

Fig. 8.

Fig. 11.

Fig. 12.

THE BRIMSTONE BUTTERFLY

PIERIDAE *Gonepteryx rhamni* L.

Although adults of this butterfly (*Figs 4, 5*) are seen in both the autumn and spring there is only one brood of this insect in the year. The adults overwinter in hollow trees, ivy clumps etc. This species is often the earliest butterfly seen on the wing in the spring as the adults may emerge from their hibernation sites on fine days in February and March.

The eggs are greenish in colour, and are laid on the midrib of the underside of leaves of the food plant. Both species of buckthorn (*Rhamnus catharticus*, common buckthorn, and *Frangula alnus*, alder buckthorn) are food plants of this insect. The larvae (*Figs 1, 2*) hatch and feed on these plants during May and June. The caterpillars then pupate on the food plant, the colour and shape of the pupa (*Fig 3*) being reminiscent of a curled leaf. The adults emerge at the end of July and in August and are seen feeding on thistles and other flowers before hibernation.

The colour of this species is different in the two sexes, males (*Fig 4*) being bright yellow, females (*Fig 5*) greenish. The butter colour of this insect may be the reason for the English name of the whole group, butterflies.

This species is found all over the British Isles wherever its food plant grows. Abroad, it ranges over much of temperate Europe and Asia, and is also found in parts of North Africa.

1 Larva 2 Larva 3 Pupa
4 Female butterfly 5 Male butterfly
Plant:Buckthorn (Rhamnus catharticus)

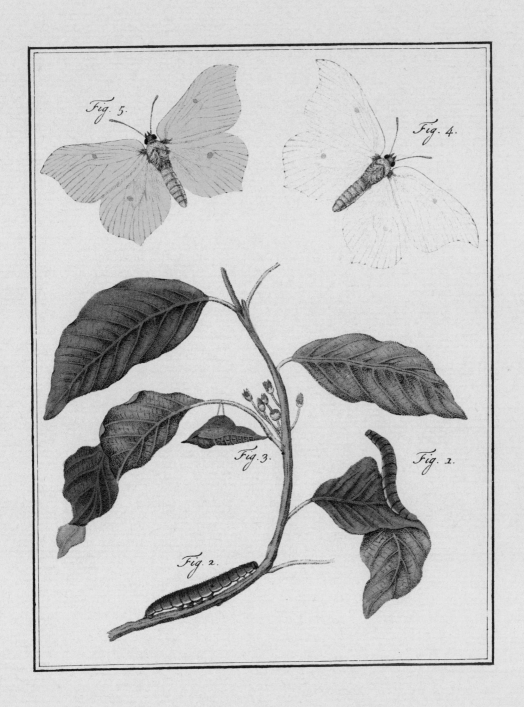

Fig. 5.

Fig. 4.

Fig. 3.

Fig. 1.

Fig. 2.

THE COMMA BUTTERFLY

NYMPHALIDAE *Polygonia c-album* L.

This species is very distinctive with the jagged indentations in the edges of its wings which so resemble a dead leaf when the butterfly closes them at rest (*Fig 4*). Also visible in this figure is the crescent shaped white mark on the underside of the hind-wings which gives this butterfly its vernacular name.

There are two broods of this insect in the year and the adults (*Figs 4, 5*) of the second brood overwinter; hence this butterfly may be seen on the wing throughout most of the warmer seasons of the year. Overwintered adults are active in early spring (April and May), first-brood adults in late June through to August and second brood adults in August and September before they seek out the hibernation sites for overwintering.

The eggs are greenish in colour and are laid, usually singly, on hop (*Humulus lupulus*) or nettle (*Urtica dioica*); they are also occasionally found on species of currant or gooseberry (*Ribes* spp.). The caterpillars (*Figs 1, 2*) feed up rapidly and become fully fed within about five weeks. They then pupate (*Fig 3*) on the food plant, to emerge as adult butterflies some ten days later.

This butterfly has a very widespread distribution outside Britain, being found across Europe and Asia as far as Japan. In this country, however, it undergoes periodic large changes in distribution and abundance and, whereas in the early part of this century it was to be found only in a few central southern counties, it has now spread again as far north as York and Cumberland.

1 Larva 2 Larva 3 Pupa
4 Adult butterfly 5 Adult butterfly
Plant:Gooseberry (Ribes uva-crispa)

THE SMALL TORTOISESHELL BUTTERFLY

NYMPHALIDAE *Aglais urticae* L.

Apart from the species of white butterflies that feed on cabbages, this is probably the butterfly most familiar to the great majority of people. There are two broods of this species in the year, with adults (*Figs 7, 8*) of the first brood on the wing in late May and June and the second brood in September and October. These second brood adults then overwinter and like those of the Brimstone butterfly may be seen on warm days in the late winter and early spring. Egg-laying however does not commence until late April or early May in most years. In the autumn it is one of the commonest visitors, along with the following three species, to garden flowers such as Michaelmas daisies and buddleia bushes in our gardens and to the thistles on waste ground.

The pale green eggs (*Figs 1, 2*) are laid in clusters on nettles (*Urtica dioica*) and the caterpillars (*Figs 3, 4*) live in conspicuous gregarious groups until they are almost fully grown, when they disperse over the food plants. These groups of caterpillars can almost completely destroy clumps of nettles and the gregarious habit may be an adaption which deters predators from eating them. The caterpillars are fully grown in four weeks or so and spin up close by (*Figs 5, 6*), to emerge after 10 days as the adult butterfly.

This species is again very widely distributed throughout most of Europe and Asia, extending to the far north.

1 Eggs, natural size
2 Egg enlarged to show detail 3 Larva
4 Larva 5 Pupa 6 Pupa
7 Adult butterfly 8 Adult butterfly
Plant:Nettle (Urtica dioica)

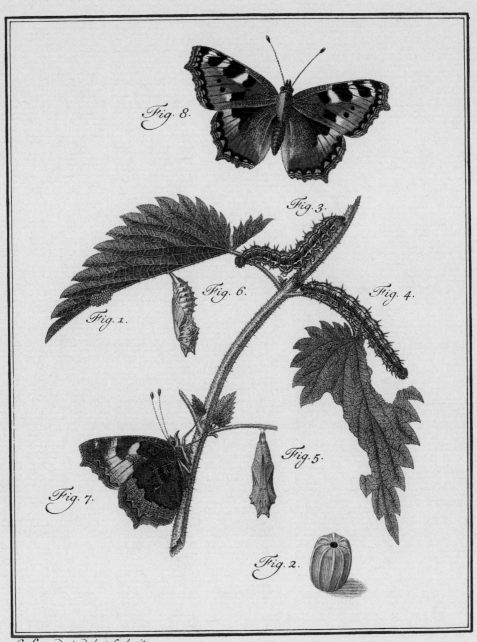

Fig. 8.

Fig. 3.

Fig. 6.

Fig. 4.

Fig. 1.

Fig. 5.

Fig. 7.

Fig. 2.

C. Sepp ad viv. del. et sculpsit.

PEACOCK BUTTERFLY

NYMPHALIDAE *Inachis io* L.

This is the second of the species that feeds on the common nettle (*Urtica dioica*) but, unlike the Small Tortoiseshell, it has only one brood in the year. It is also much more in evidence in the autumn feeding in the garden than it is in the spring after the winter has taken its toll of the hibernating adults.

The overwintered adults (*Figs 6, 7*) emerge from their hibernation site in hollow trees and outhouses from late March to early May and lay their eggs in batches on the undersides of nettle leaves. The caterpillars feed gregariously throughout their life and are, again, very conspicuous, not only because of their dark colouring, but also because of the extensive damage that they do to the plants on which they are feeding.

The pupae (*Fig 5*) are found on the food plant and the adult butterflies emerge from late July to September and feed up on flowers and rotting fruit until hibernation sets in. This species is very widely distributed both within Britain and across Europe and Asia as far as Japan.

1 Eggs laid on leaf 2 Eggs, natural size
3 Egg enlarged to show detail 4 Larva
5 Pupae 6 Adult butterfly
7 Adult butterfly
Plant:Nettle (Urtica dioica)

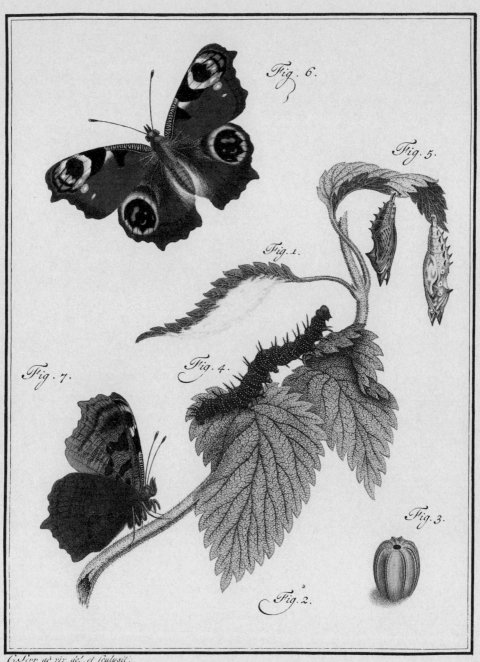

Fig. 6.

Fig. 5.

Fig. 1.

Fig. 7.

Fig. 4.

Fig. 3.

Fig. 2.

C. Sepp ad vir. del. et sculpsit.

THE PAINTED LADY BUTTERFLY

NYMPHALIDAE *Vanessa cardui* L.

This butterfly probably has the widest distribution of any insect in this group and is found almost throughout the world. In much of Europe it does not survive through the winter, but it is a strong flier and the adults (*Figs 7, 8, 9*) migrate northwards in the spring from the insects' winter breeding areas in North Africa.

These spring immigrants arrive in Britain in May and June, and lay their eggs (*Figs 1, 2*) on various species of thistles, and occasionally on nettles and other plants. The caterpillars (*Fig 3*) hatch and feed up quickly before spinning up (*Fig 4*) on the food plant or elsewhere nearby. The pupae (*Figs 5, 6*) produce butterflies seen on the wing in our gardens and fields during the late summer and autumn (August to October). These do not hibernate like those of the Peacock butterfly, present at the same time, but try to produce another brood of larvae before being killed off by severe weather.

This species and the Red Admiral (*pp 48, 49*) are not only strong migrants but often move at night as well as during the day and are sometimes taken at light-traps set up to capture night-flying moths.

1 Egg, natural size
2 Egg enlarged to show detail 3 Larva
4 Pre-pupa 5 Pupa
Plant:Thistle (Cirsium sp)

Fig. 5.

Fig. 3.

Fig. 4.

Fig. 2.

Fig. 1.

THE PAINTED LADY BUTTERFLY

NYMPHALIDAE *Vanessa cardui* L.

(continued)

6 Pupa 7 Adult butterfly
8 Adult butterfly 9 Adult butterfly
Plant:Mallow (Malva sylvestris)

Fig. 6.

Fig. 8.

Fig. 7.

Fig. 9.

THE RED ADMIRAL BUTTERFLY

NYMPHALIDAE *Vanessa atalanta* L.

The life history of this species is very like that of the Painted Lady, but it does not have quite such a wide distribution. It is found, however, on all the northern continents and as an introduced species in New Zealand. It is probably very rare for adults of this species to overwinter successfully in Britain and other parts of northern Europe; hence the population is almost entirely derived from immigrants arriving on our shores in the spring. This dependence on immigration means that the number of adults (*Figs 10, 11*) observed in each year varies according to the number of those spring immigrants, and to their success at breeding in our uncertain summers.

The eggs (*Figs 1, 2, 3*) are laid singly on young nettle leaves. The caterpillars (*Figs 4, 5*) spin several leaves together to form a sort of shelter and grow quickly, becoming fully fed (*Fig 6*) in about four or five weeks. They then spin up (*Figs 7, 8*) to form the pupa (*Fig 9*), usually inside the larval shelter.

The adult butterflies emerge in August and September and are found feeding at flowers such as buddleia. This species is also attracted to trees that have been attacked by Goat moth (*Cossus cossus*), the larvae of which bore into the trunk. Many butterflies may be seen around these trees at one time, feeding on the exudation from the injury.

1 *Egg, natural size*
2 *Egg enlarged to show detail*
3 *Shelter which the larva builds for itself*
4-5 *Larvae 6-8 Stages in pupation*
9 *Pupa 10 Adult butterfly*
11 *Adult butterfly*
Plant: Nettle (Urtica dioica)

Fig. 3.

Fig. 4.

Fig. 8.

Fig. 9.

d

b

c

Fig. 5.

Fig. 6.

Fig. 7.

a

Fig. 11.

Fig. 10.

Fig. 2.

Fig. 1.

THE DARK GREEN FRITILLARY BUTTERFLY

NYMPHALIDAE *Mesoacidalia aglaja* L.

This species is perhaps one of the commonest of the group of orange and brown butterflies known as fritillaries. The adults (*Figs 4, 5*) fly in July and August and are found in many places in Britain where there are open areas of rough grassland and waste ground, although the species is perhaps most frequent near to the coast.

It is a strong fast flier but can often be seen at rest on the taller species of thistles. The yellow eggs are laid on violets and soon darken in colour and hatch. The young caterpillar does not feed but soon goes into hibernation. Occasionally, in northern areas, the eggs may not hatch in the autumn but may overwinter in this stage.

In the spring the caterpillar (*Figs 1, 2*) commences feeding on the leaves of dog violet (*Viola canina*) or occasionally pansies (*Viola* spp.). It becomes fully grown by late June and moults to the pupal stage (*Fig 3*) after spinning a tent-like arrangement of leaves in which to hide. The adult insect emerges in July.

Despite its apparent liking for coastal habitats in this country, it has a wide distribution abroad, covering much of Europe and North Africa and extending across Asia to Japan.

1 Larva 2 Larva 3 Pupa
4 Adult Butterfly 5 Adult butterfly
Plant: Dandelion (Taraxacum officinale)

Fig. 5.

Fig. 4.

Fig. 2.

Fig. 1.

Fig. 3.

THE MARSH FRITILLARY BUTTERFLY

NYMPHALIDAE *Euphydryas aurinia* Rot.

This species is uncommon but can be found in suitable habitats through-out most of southern Britain. The adult (*Figs 4, 5*) is seen on the wing in May and June in damp meadows and heathlands.

The eggs are light yellow in colour and are laid in large batches, usually on devils bit scabious (*Succisa pratensis*), and occasionally on ribwort plantain (*Plantago lanceolata*).

The larvae hatch in July and spin a dense silken web as a shelter amongst the leaves of the food plant. They retire to these shelters at night and during dull weather—they are daytime feeders only. They also pass the winter as partly grown caterpillars in these webs. In March they recommence feeding and soon become fully grown larvae (*Figs 1, 2*). The pupa is attached to a silk pad spun on the underside of a leaf (*Fig 3*) and the adult butterfly emerges in late May.

This species is found throughout much of Europe, North Africa and Asia as far as Korea.

1 Larva 2 Larva 3 Pupa
4 Adult butterfly 5 Adult butterfly
Plant: Ribwort plantain (Plantago
lanceolata)

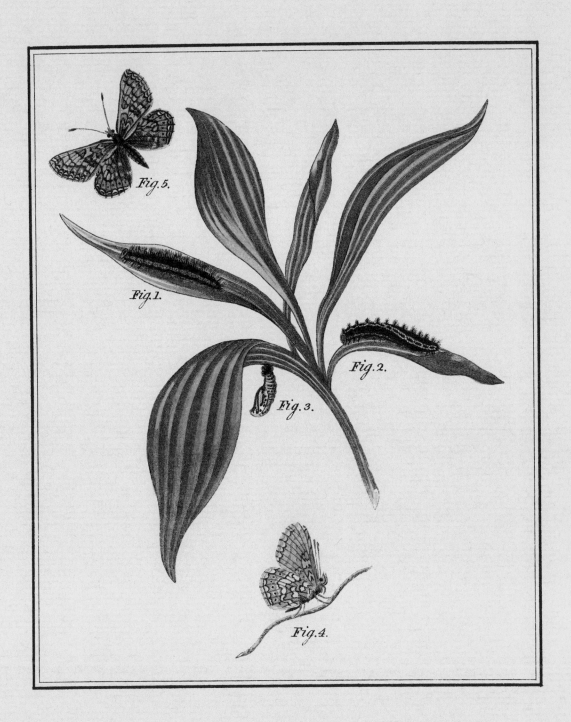

Fig. 5.

Fig. 1.

Fig. 2.

Fig. 3.

Fig. 4.

THE GLANVILLE FRITILLARY BUTTERFLY

NYMPHALIDAE *Melitaea cinxia* L.

This species has rather a southern distribution, although it occurs locally throughout much of Europe including Scandinavia. In Britain, however, it is only found in the Isle of Wight and in the Channel Islands.

The adult butterfly (*Figs 4, 5*) is on the wing in May and June. The eggs are laid in batches on the undersides of the food plant's leaves. The preferred plants in this country appear to be the sea plantain (*Plantago maritima*) and the ribwort plantain (*Plantago lanceolata*) amongst which the young larvae spin their silken shelters. Here they hibernate for the winter as half-grown caterpillars. These larvae (*Figs 1, 2*) commence feeding again in early spring, completing their growth by mid-April. They then pupate (*Fig 3*) on the food plant and emerge as adults in about three weeks.

1 Larva 2 Larva 3 Pupa
4 Adult butterfly 5 Adult butterfly
Plant: Ribwort plantain (Plantago
lanceolata)

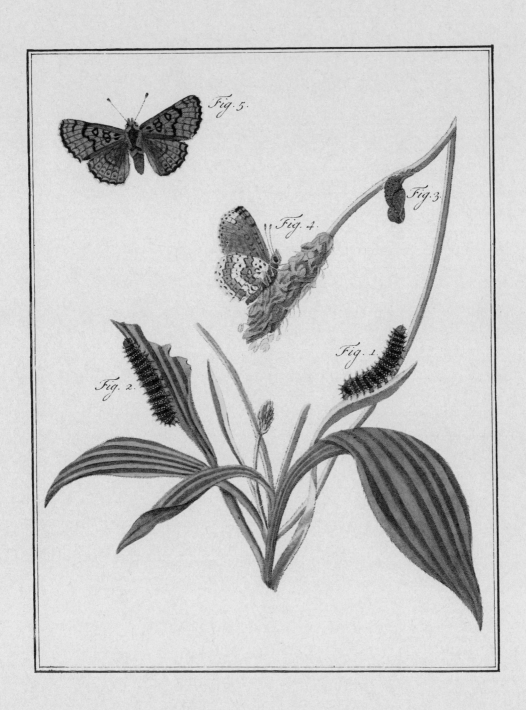

Fig. 5.

Fig. 3.

Fig. 4.

Fig. 1.

Fig. 2.

THE SPECKLED WOOD BUTTERFLY

Satyridae *Pararge aegeria* L.

This species, unlike most other butterflies, frequents shady woods and lanes. It is found as an adult (*Figs 6, 7*) throughout much of the year and has several broods in the year; these overlap each other so that all stages of this species may be found at any one time.

Both pupae and partly grown larvae are known to overwinter, the pupae giving rise to the adults seen on the wing in April and May while the overwintered larvae produce adults in June. Eggs laid in May produce adults in July, while those laid in July and June produce the adults in the autumn.

The eggs (*Figs 1, 2*) are laid on a number of species of grass, particularly couchgrass (*Agropyron repens*), meadowgrass (*Poa* spp.) and false-brome (*Brachypodium sylvaticum*). The larvae (*Fig 3*), which are well camouflaged by virtue of their green colour and lighter stripes, feed up rapidly and then pupate (*Figs 4, 5*) suspended from a silken pad some five weeks after hatching.

The whole life cycle from egg to adult takes about seven weeks in the summer so that often at least three broods are produced in the year from the overwintering pupae and two from the larvae.

This butterfly is widespread in suitable habitats throughout Europe, western Asia and North Africa.

1 Egg, natural size
2 Egg enlarged to show detail 3 Larva
4 Pupa 5 Pupa 6 Adult butterfly
7 Adult butterfly
Plant:Bentgrass (Agrostis sp)

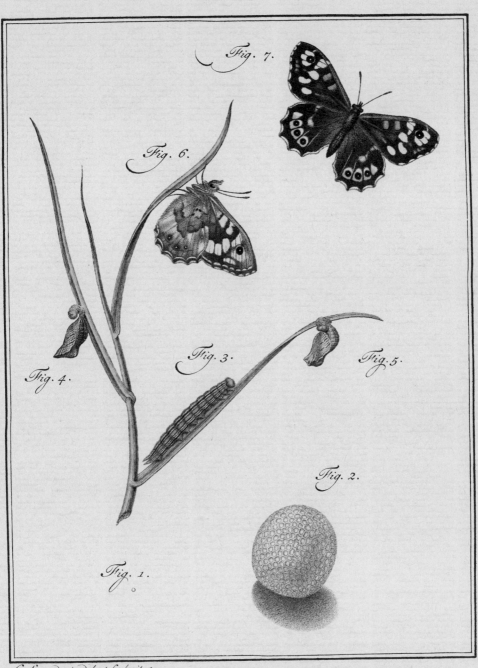

Fig. 7.

Fig. 6.

Fig. 4.

Fig. 3.

Fig. 5.

Fig. 2.

Fig. 1.

THE WALL BROWN BUTTERFLY

SATYRIDAE *Lasiommata megera* L.

This species, which is common throughout most of England and Wales, though more local in Scotland, gets its English name from its habit of basking on walls and stones. It has two generations each year, adults (*Figs 7, 8, 9*) being found in May and June and again in late July, August and September.

The females (*Fig 9*) are generally somewhat more brightly marked than the males (*Fig 8*), which have a conspicuous large brand mark making the central band on the fore-wings much thicker and more obvious.

The eggs (*Figs 1, 2*) are laid singly and the caterpillars (*Figs 3, 4, 5*) feed up on a variety of grasses, although they seem to like meadowgrass (*Poa* spp.) and cocksfoot (*Dactylis glomerata*) best. The pupae (*Figs 5, 7*) are found on grass stems, the green form being the commonest. The insects in the autumn generation hibernate as partly grown larvae and feed up in the spring to pupate in April.

This butterfly is found largely in the rough grasslands along woodland edges and often visits gardens. It is common through much of Europe, North Africa and parts of the Middle East.

1 Egg, natural size
2 Egg enlarged to show detail 3 Larva
4 Pre-pupa 5 Pupa
Plant:Cocksfoot (Dactylis glomerata)

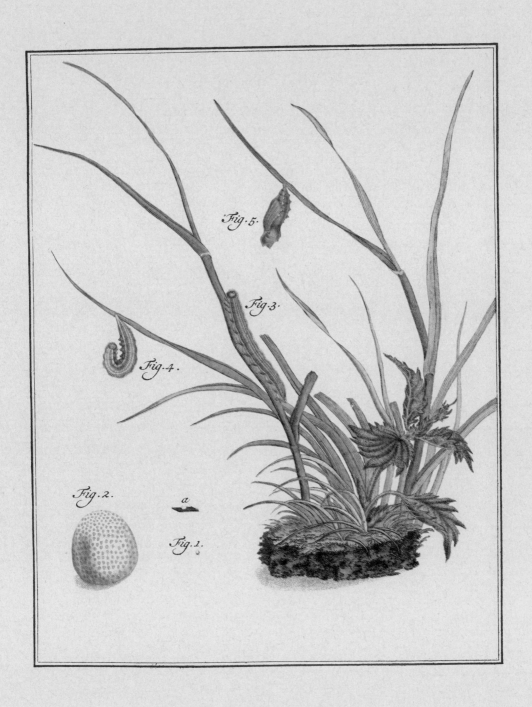

Fig. 5.

Fig. 3.

Fig. 4.

Fig. 2.

a

Fig. 1.

THE WALL BROWN BUTTERFLY

Satyridae *Lasiommata megera* L.

(continued)

6 Larva 7 Pupa 8 Male butterfly
9 Female butterfly 10 Butterfly at rest
Plant:Cocksfoot (Dactylis glomerata)

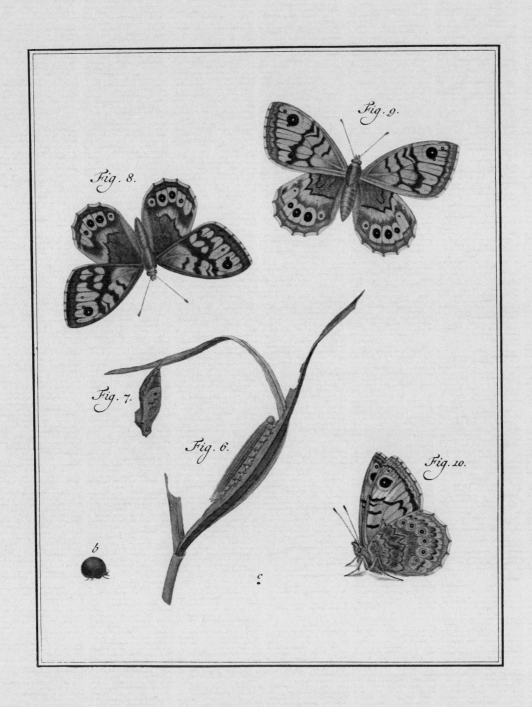

Fig. 9.

Fig. 8.

Fig. 7.

Fig. 6.

Fig. 10.

b

c

THE MEADOW BROWN BUTTERFLY

SATYRIDAE *Maniola jurtina* L.

This species must be our most familiar meadow butterfly, being found on the wing in most grassy places in fields and waste ground from June to September. The males and females (*Figs 5, 6*) differ in size and the amount of light coloration on the wings. The adult butterflies illustrated are both females; they are slightly larger than the males, which have a much smaller amount of the light coloration, usually confined to a small patch immediately surrounding the eye spots on the fore-wings.

Despite its long flight period this species has only one brood in the year. The eggs (*Figs 1, 2*) are laid on grass blades and the caterpillars (*Fig 3*) hatch and feed until late summer, after which they go into hibernation. The host plant may be one of a number of species of grass, but annual meadowgrass (*Poa annua*) and cocksfoot (*Dactylis glomerata*) seem to be preferred species. The caterpillars commence feeding again early in the spring and complete their development by the end of May. They are strictly night-feeding, coming up from the lower layers in the sward where they hide during the day. This sort of feeding pattern is common in many grassland insects. When fully fed, the caterpillars spin up on a grass blade and the adults emerge from these pupae (*Fig 4*) in June and July.

This species of butterfly is common in North Africa, parts of the Middle East and throughout Europe as far as the Urals.

1 *Egg, natural size*
2 *Egg enlarged to show detail 3 Larva*
4 *Pupa 5 Male butterfly*
6 *Female butterfly*
Plant: Bentgrass (Agrostis sp)

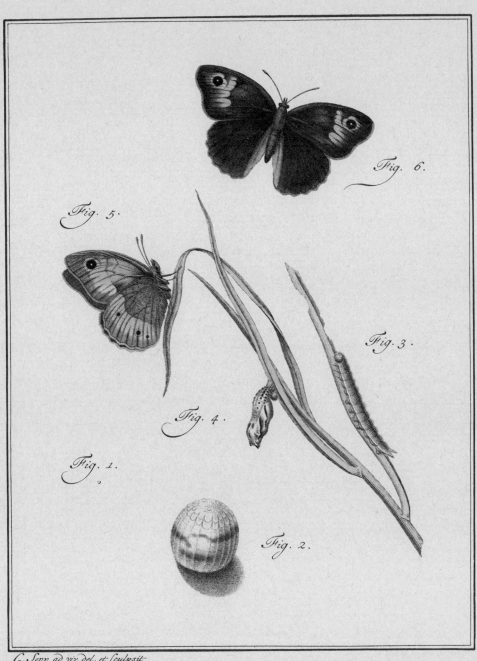

Fig. 6.

Fig. 5.

Fig. 3.

Fig. 4.

Fig. 1.

Fig. 2.

THE GATEKEEPER BUTTERFLY

SATYRIDAE *Pyronia tithonus* L.

This is another common grassland butterfly, at least in the southern part of the country. The adults (*Figs 7, 8*) are often found around hedgerows and woodland edges feeding on the flowers of bramble (*Rubus* spp.), wood sage (*Teucrium scorodonia*) and wild marjoram (*Origanum vulgare*). They are on the wing from July to September and there is only one brood each year. The females (*Fig 8*) are larger and more brightly coloured than the males (*Fig 6*) which have a large brownish band in the middle of the fore-wings.

The eggs (*Figs 1, 2*) are laid on various grasses but especially on annual meadowgrass (*Poa annua*), wood millet (*Millium effusum*) and couchgrass (*Agropyron repens*). The caterpillars (*Figs 3, 4*) overwinter in the middle instar and feed up again in the following spring. They too are a strictly nocturnal species. The pupae (*Fig 5*) are found on the grasses.

This species occurs in western Europe and Asia Minor but is less widespread in the areas around the Mediterranean.

1 Egg, natural size
2 Egg enlarged to show detail 3 Larva
4 Larva 5 Pupa 6 Male butterfly
7 Adult butterfly at rest
8 Female butterfly
Plant:Bentgrass (Agrostis sp)

Fig. 6.

Fig. 8.

Fig. 3.

Fig. 5.

Fig. 4.

Fig. 2.

Fig. 1.

Fig. 7.

THE RINGLET BUTTERFLY

SATYRIDAE *Aphantopus hyperanthus* L.

This dark butterfly is common in much of Britain on woodland edges and grassy wastelands. The adults (*Figs 5, 6, 7, 8*) are on the wing in July and August. Generally the females (*Figs 7, 8*) are larger than the males (*Figs 5, 6*), with more distinct spots on the upper surface of the wings and larger ones on the underside.

The eggs (*Figs 1, 2*) are laid on various species of grass, with once again annual meadowgrass (*Poa annua*) apparently the preferred species. The caterpillars (*Fig 3*) feed up slowly, overwintering as half-grown larvae, and eventually reach full size in the following June. The pupae (*Fig 4*) are found in the lower layers of grass tussocks and emerge after three or four weeks.

This species is widespread in Europe, especially western and central areas, but is largely absent from the Mediterranean region. It is also found through central and northern Asia as far as Japan.

1 Egg, natural size
2 Egg enlarged to show detail 3 Larva
4 Pupa 5-6 Male butterflies
7-8 Female butterflies
Plant: Bentgrass (Agrostis sp)

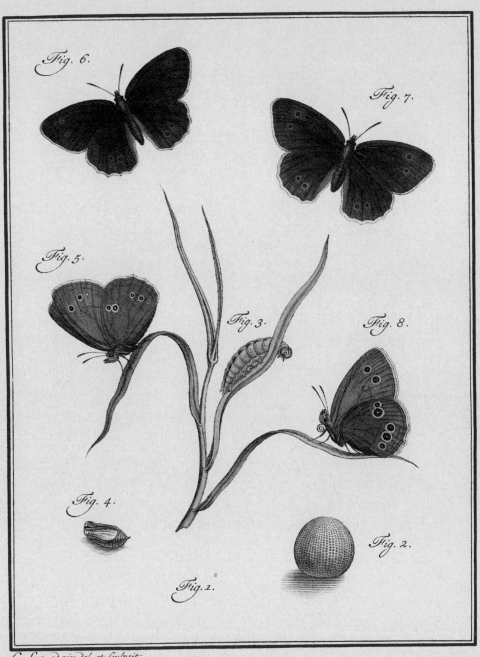

Fig. 6.

Fig. 7.

Fig. 5.

Fig. 3.

Fig. 8.

Fig. 4.

Fig. 2.

Fig. 1.

C. Sepp ad vir. del. et sculpsit.

THE SMALL HEATH BUTTERFLY

SATYRIDAE *Coenonympha pamphilus* L.

This small yellowish butterfly is very common throughout the summer in almost every patch of grass or heathland in the country. The adults (*Figs 5, 6, 7*) are found from May to September and, as with the Speckled Wood, there is a succession of broods throughout the summer. The males (*Fig 5*) are similar to the females (*Fig 6*) but are usually smaller and with a more strongly marked greyish border to the wings.

The eggs (*Figs 1, 2*) are laid singly on many grasses, though annual meadow grass (*Poa annua*) and moorgrass (*Nardus stricta*) seem to be preferred. After hatching, some of the caterpillars (*Fig 3*) feed up quickly, becoming fully fed in some four or five weeks while the rest feed up much more slowly, overwinter as partly grown larvae and complete their growth in the following spring. The pupae (*Fig 4*) are found attached to the grass and the adults emerge after ten to fourteen days.

This species is found throughout Europe and North Africa, the Middle East and in western and central Asia.

1 Egg, natural size
2 Egg enlarged to show detail 3 Larva
4 Pupa 5 Male butterfly
6 Female butterfly
7 Adult butterfly at rest
Plant:Cocksfoot (Dactylis glomerata)

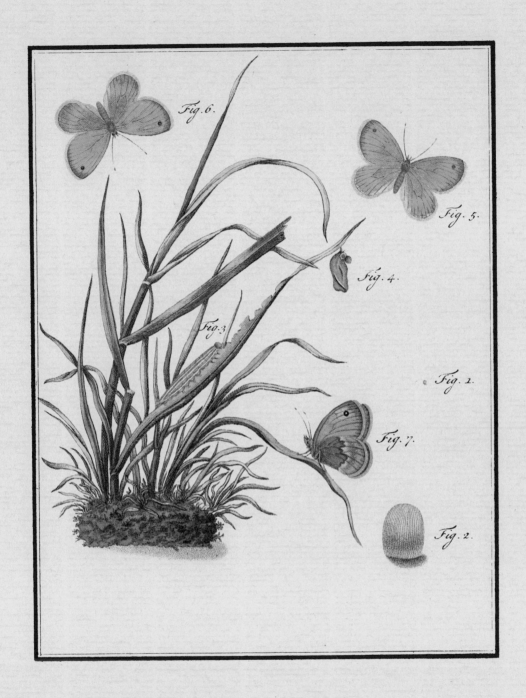

Fig. 6.

Fig. 5.

Fig. 4.

Fig. 3.

Fig. 1.

Fig. 7.

Fig. 2.

THE PURPLE HAIRSTREAK BUTTERFLY

LYCAENIDAE *Quercusia quercus* L.

This is the most widespread representative of the group of butterflies known as hairstreaks because of the fine white markings on the under-surface of the wings of many species. In Britain most of the other species of this group are very local woodland butterflies.

This pretty species is found in oak woods and is on the wing in July and August. Much of the time it flies high in the trees but in hot weather it can be found basking on areas of bare ground, such as paths and driveways. It can also often be seen feeding on the flowers of the sweet chestnut (*Castanea sativa*).

The eggs are laid singly on the twigs of oak (*Quercus* spp.) and they overwinter before hatching in the following spring. The slug-like cater-pillars (*Figs 1, 2*) are fully fed by late May or early June and usually descend to the ground to pupate (*Fig 3*).

The adults (*Figs 4, 5, 6*) are strongly sexually dimorphic, the males (*Fig 6*) being almost entirely bright metallic purple in colour, while the females (*Fig 4*) have only a patch of this coloration on the fore-wings.

This species is found in suitable habitats throughout much of Britain and across Europe and North Africa.

1 Larva 2 Larva 3 Pupa
4 Female butterfly
5 Adult butterfly at rest 6 Male butterfly
Plant:Oak (Quercus sp)

Fig. 6.

Fig. 4.

Fig. 1.

Fig. 2.

Fig. 3.

Fig. 5.

THE SILVER-STUDDED BLUE BUTTERFLY

Lycaenidae *Plebejus argus* L.

There are several other species of blue butterfly, some of which are commoner than the Silver-studded Blue. These however were not illustrated by Sepp, and so were not available for inclusion in the present volume. The commonest of these species is the Common Blue (*Polyommatus icarus*) which is similar in coloration but slightly larger than the species illustrated here and lacks the silver spots on the under-wings.

The males of the Silver-studded Blue (*Figs 3, 4*) are brightly coloured while the females (*Figs 5, 6*) are much duller and darker in colour. The silver spots which give the species its English name are found in the eyespots on the undersurface of the hind-wings. (*Figs 3, 5*).

The porcelain-like eggs are laid singly on a variety of leguminous plants and sometimes on other plants. They overwinter and hatch the following April. The larvae (*Fig 1*) feed on the flowers and seed pods of, for example, gorse (*Ulex* spp.), broom (*Sarothamnus scoparius*), bird's-foot trefoil (*Lotus corniculatus*) or heather (*Calluna vulgaris*). They feed up in a few weeks and pupate (*Fig 2*) in June, the adults emerging some ten days later.

This species is common in England and Wales on grassy heaths and moors, but is very restricted in its distribution in Scotland. Abroad, it is found throughout Europe and the temperate parts of Asia as far as Korea and Japan.

1 Larva 2 Pupa 3 Male butterfly
4 Male butterfly 5 Female butterfly
6 Female butterfly
Plant:Broom (Sarothamnus scoparius)

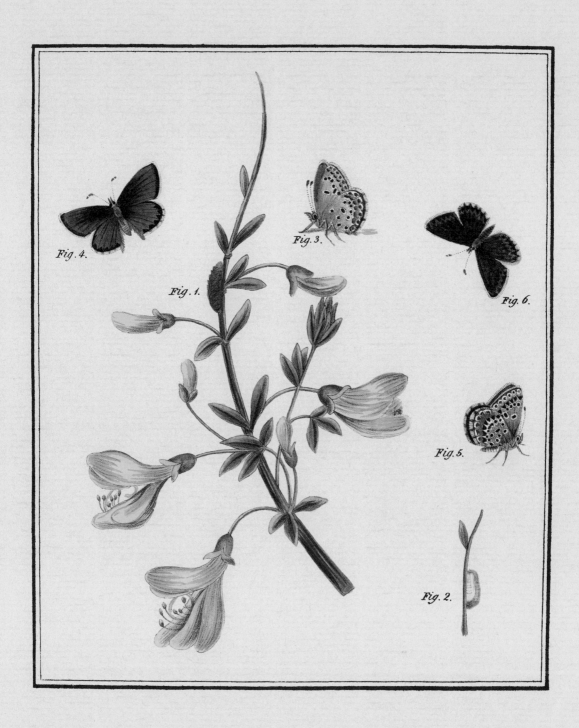

Fig. 4.

Fig. 3.

Fig. 6.

Fig. 1.

Fig. 5.

Fig. 2.

THE LIME HAWKMOTH

SPHINGIDÀE *Mimas tiliae* L.

This handsome insect is probably one of the more familiar hawkmoths to town dwellers in the south where lime trees (*Tilia*) are widely planted for street shade.

The adult moth (*Figs 4, 5*) emerges in May and June and may be found sitting on the trunks of the trees where the larvae feed or on fences nearby. These specimens are often newly emerged moths; they usually emerge from the pupa in the late afternoon. The colour pattern of the adult is not very variable but occasionally specimens may be found with the dark band on the fore-wings broken or reduced to a central spot.

Most eggs (*Fig 1*) are laid, usually singly, on the leaves of lime or elm but some are occasionally found on a number of other tree and shrub species. The eggs hatch in ten days or so and the young larvae begin to feed. At this stage the colouring is less intense and they usually lack the yellow stripes on the sides of the body. The caterpiller (*Fig 2*) becomes fully grown by mid to late August and descends from the tree to pupate. The fully grown caterpillar burrows a few inches into the ground and constructs a fragile earthen chamber, usually within a few feet of the tree on which it fed. Within the cell it moults into the pupa (*Fig 3*) which remains in the ground until the adult emerges in the following spring.

This moth is common in the southern part of Britain wherever lime trees are abundant. Abroad it is found in large parts of Europe wherever its food plant grows.

1 Eggs 2 Larva 3 Pupa 4 Female moth
5 Male moth
Plant:Lime (Tilia sp)

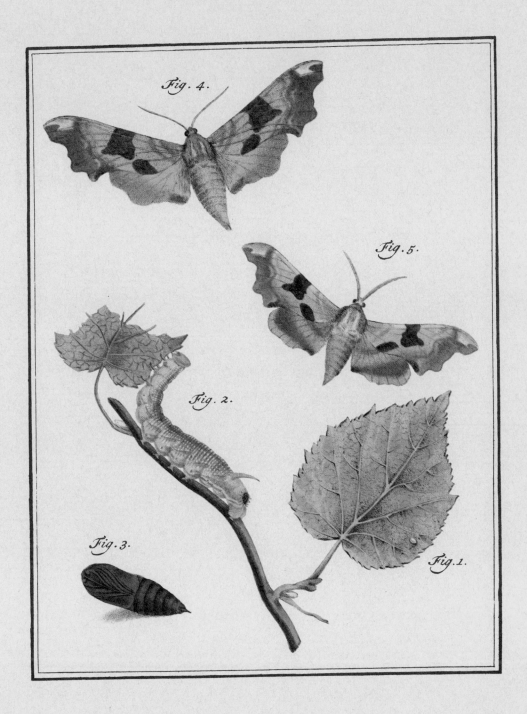

Fig. 4.

Fig. 5.

Fig. 2.

Fig. 3.

Fig. 1.

THE POPLAR HAWKMOTH

SPHINGIDAE *Laothoë populi* L.

This is the commonest of our resident hawkmoths, and is found over the whole of the country wherever willows and poplars grow.

There are generally two broods of this moth each year, but in cool seasons there may be only one. The adults (*Figs 8, 9*) are normally on the wing in May and June and again in late July and August.

The eggs (*Figs 1, 2*) are laid singly on a range of tree species, but mostly on poplar (*Populus* spp.), willows and sallows (*Salix* spp.) or birch (*Betula* spp.). The caterpillars (*Figs 3, 4, 5*) hatch in about ten days and feed up rapidly. They become conspicuous and can be easily found in the later stages because of the large amount of foliage they consume. When it is fully grown the caterpillar descends to the ground and constructs a cell several inches underground in which to pupate. The pupae (*Fig 7*) of the first brood hatch in a few weeks to produce adult moths while those of the second brood overwinter as pupae.

The adult moth comes readily to a light source and may be found sitting around street lamps or resting on fences and tree trunks nearby.

This species is found throughout Europe as far north as the Arctic Circle.

1 Egg, natural size
2 Egg enlarged to show detail
3-6 Larvae 7 Pupa 8 Male moth
9 Female moth
Plant:Willow (Salix sp)

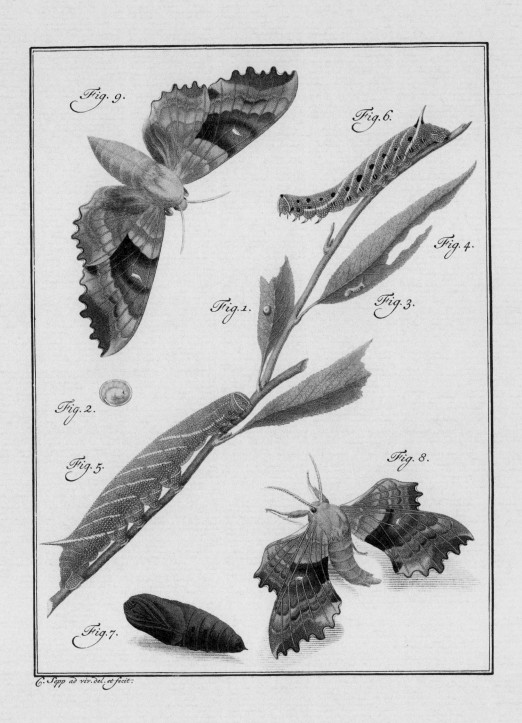

Fig. 9.

Fig. 6.

Fig. 4.

Fig. 1.

Fig. 3.

Fig. 2.

Fig. 5.

Fig. 8.

Fig. 7.

C. Sepp ad viv. del. et fecit.

THE EYED HAWKMOTH

SPHINGIDAE *Smerinthus ocellata* L.

This species, with its spectacular eyed hind-wings, is fairly common in southern England but much less common in the north. When the moth is at rest the eyespots are covered by the fore-wings (*Fig 6*), but when the wings are opened they give the impression of a much larger animal, thus, it is said, scaring predatory birds sufficiently to allow the moth to escape.

This moth usually has only one brood. The adults (*Figs 6, 7*) are on the wing in late May and June, but in some years, when adults may appear early in May, there can be a partial second brood on the wing in July and August.

The eggs (*Fig 1*) are laid singly or in pairs on willows and sallows (*Salix* spp.) or on apple (*Malus* spp.). The caterpillars (*Figs 2, 3, 4*) feed up quickly, taking some six to eight weeks to reach full size. They then pupate underground and overwinter as a pupa (*Fig 5*), but in warm years pupae formed in June may emerge in July.

1 Egg 2-4 Larvae 5 Pupa
6 Female moth 7 Male moth
Plant:Willow (Salix sp)

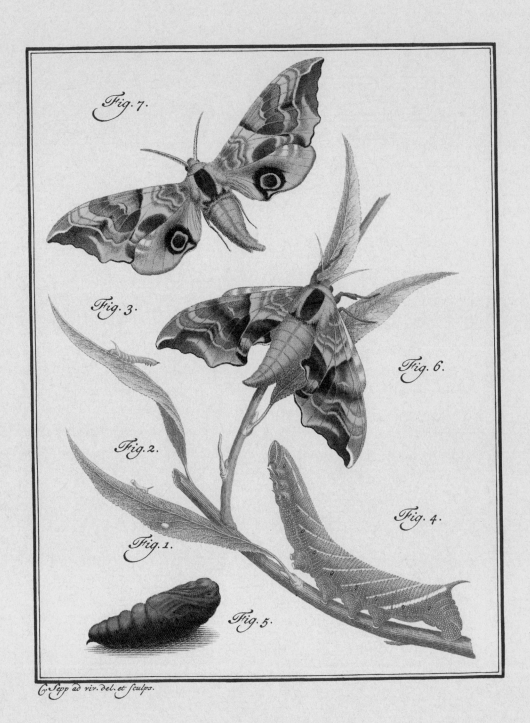

Fig. 7.

Fig. 3.

Fig. 6.

Fig. 2.

Fig. 4.

Fig. 1.

Fig. 5.

C. Sepp ad vir. del. et sculps.

THE DEATH'S HEAD HAWKMOTH

Sᴘʜɪɴɢɪᴅᴀᴇ *Acherontia atropos* L.

This species and the Convolvulus Hawkmoth (pages 84-87) are probably the largest and most spectacular moths to reach Britain. They are rather irregular in their occurrences as, like the Red Admiral and the Painted Lady amongst the butterflies, they do not survive the winter in this country. The number of specimens found in any year is therefore dependent both on the weather and on the quantity of initial immigrants reaching our shores from the continent or North Africa. Like all of the strong-flying species they are very widely distributed abroad, being found in most of Europe, Asia and Africa.

The adults of the Death's-head Hawkmoth (*Fig 4*) are unusual amongst the Lepidoptera in that they are able to emit a high-pitched squeak, especially when handled. This squeak is said to resemble the piping of the queen bee and to be used by the moth when raiding nests of bees for honey.

Although in southern Europe this species has several broods in the year, and indeed they have been found in most months of the year in Britain, there are two definite peaks in this country. The first of these is in May and June when the insects first arrive and the second in September and October is probably in part the result of insects completing their development from eggs laid by the spring immigrants.

The eggs are laid singly on a variety of plants, most commonly potato (*Solanum tuberosum*), bittersweet (*S. dulcamara*) and snowberry (*Symphoricarpos rivularis*). The large caterpillars (*Fig 1*) are perhaps more frequently seen than the adults as they, and the subterranean pupae (*Figs 2, 3*), are often turned up during the potato harvest.

1 Larva
Plant: Potato (Solanum tuberosum)

Fig. 1.

THE DEATH'S HEAD HAWKMOTH

SPHINGIDAE *Acherontia atropos* L.

(continued)

2 Pupa 3 Pupa 4 Adult moth

Fig. 4.

Fig. 3.

Fig. 2.

THE CONVOLVULUS HAWKMOTH

Sphingidae *Herse convolvuli* L.

This large species occurs somewhat more regularly than does the Death's-head Hawkmoth. The adults (*Figs 5, 6*) have an extremely long tongue and feed at dusk from many flowers, although they are particularly attracted to flowers of petunia and similar tubular blossoms. They do not settle to feed but hover on the wing in front of each bloom in turn and probe into it with the long proboscis.

The eggs (*Fig 1*) are laid on various species of bindweed (*Convolvulus* spp.), the favourite apparently being the lesser bindweed (*C. arvensis*). The large larvae (*Fig 2*) feed up rapidly and pupate several inches underground. The pupa (*Figs 3, 4*) is unusual in that the long proboscis is encased in a long projection at the front end of the pupa, looking rather like the handle of a jug.

The adults have been found in many months of the year; the peak is certainly in August and September, but a number have been reported in May and June. How many of these are new arrivals and how many have successfully emerged from eggs laid by spring immigrants is not clear, but the species is certainly able to breed here during the summer months.

Eggs hatching in June produce, in July, fully grown caterpillars which after pupation become adult moths in September; the larvae and pupae from this brood are normally killed by bad weather in October and November.

1 Egg 2 Larva 3 Pupa
Plant: Field bindweed (Convolvulus arvensis)

Fig. 2.

Fig. 1.

Fig. 3.

THE CONVOLVULUS HAWKMOTH

SPHINGIDAE *Herse convolvuli* L.

(continued)

4 Pupa 5-6 Adult moths

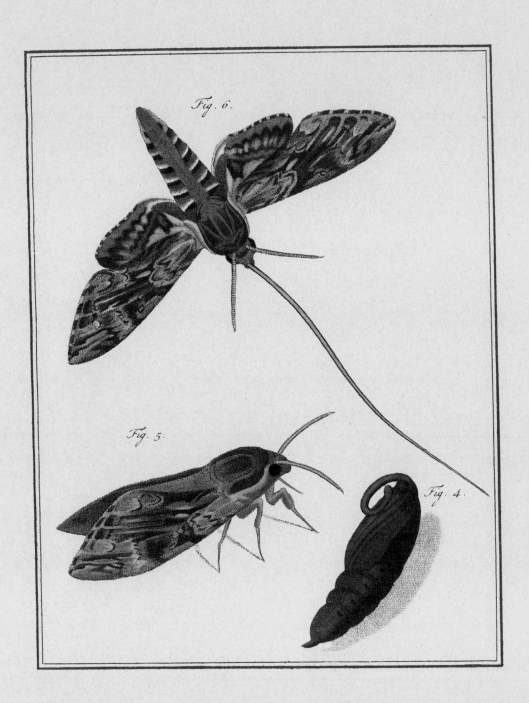

Fig. 6.

Fig. 5.

Fig. 4.

THE PRIVET HAWKMOTH

SPHINGIDAE *Sphinx ligustri* L.

This large species is fairly common and widely distributed in the southern part of the country, but it is local in the north and rare in Scotland. It is associated with gardens, as its main food plant, privet (*Ligustrum vulgare*) is widely planted as a hedge. It has been found feeding on a number of other species of shrubs and trees, for example, lilac (*Syringa vulgaris*), ash (*Fraxinus excelsior*) and mock orange (*Philadelphus coronarius*).

The adults (*page 91, Figs 2, 3*) are found in June and July and there is only one generation in the year. The eggs (*page 89, Figs 1, 2*) are found singly on the leaves of the food plant and the caterpillars (*page 89, Figs 3, 4, 5*) feed up quickly, becoming fully grown in about five weeks. They then descend to the soil and pupate (*page 91, Fig 1*) in an earthen cell several inches underground. The adult usually emerges in the following summer, but may lay over for more than one year before emerging.

This species is found throughout much of Europe and Asia as far as Japan.

1 Eggs, natural size
2 Eggs enlarged to show detail
3-5 Larvae
Plant: Privet (Ligustrum vulgare)

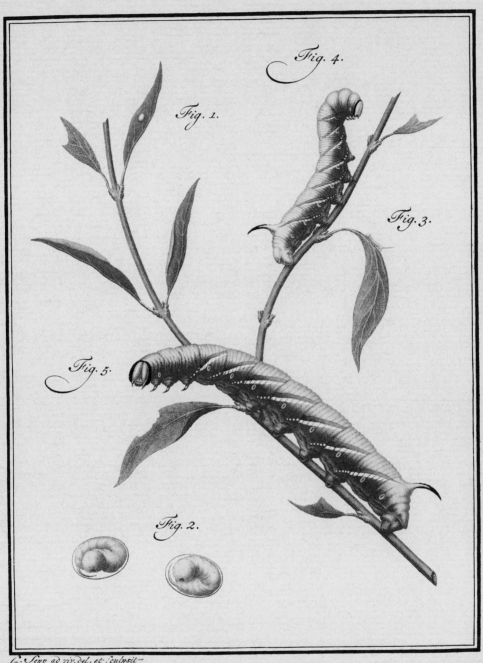

Fig. 4.

Fig. 1.

Fig. 3.

Fig. 5.

Fig. 2.

C. Sepp ad viv. del. et sculpsit

THE PRIVET HAWKMOTH

Sphingidae *Sphinx ligustri* L.

(continued)

1 Pupa 2 Adult moth 3 Adult moth
Plant:Privet (Ligustrum vulgare)

Fig. 3.

Fig. 2.

Fig. 1.

C. Sepp ad viv. del. et sculpsit.

THE PINE HAWKMOTH

SPHINGIDAE *Hyloicus pinastri* L.

This species is widespread in Britain, but very local in occurrence. Recently it may have become much more common than previously due to the activities of our foresters in planting large stands of conifers.

The adults (*Figs 8, 9*) are on the wing in July and August and may be seen feeding at honeysuckle flowers in the evening. The greyish coloration of the fore-wings provides an excellent camouflage when the moth is at rest on the trunk of a pine tree.

The eggs (*Figs 1, 2*) are laid singly on the needles of various species of pine (*Pinus* spp.), and the larvae hatch and feed up in about two months. At first they are striped (*Figs 3, 4*) to match the colour and size of the needles but in later instars (*Fig 6*) they are too large for this type of concealment and they become much more brightly coloured. The effect of this is to break up their outline sufficiently to conceal them.

They descend the tree to pupate (*Fig 7*) in October and emerge in the following summer.

This species is found in northern and central Europe and Asia.

1 Egg, natural size
2 Egg enlarged to show young larva inside
3-4 Larvae 5 Detail of larval horn
6 Fully fed larva 7 Pupa
8 Female moth 9 Male moth
Plant:Pine (Pinus sp)

Fig. 9.

Fig. 1.

Fig. 3.

Fig. 6.

Fig. 2.

Fig. 4.

Fig. 8.

Fig. 5.

Fig. 7.

THE ELEPHANT HAWKMOTH

SPHINGIDAE *Deilephila elepnor* L.

This pretty pink and brown species is one of our commoner hawkmoths and is found as far north as central Scotland. The adult moths (*Figs 6, 7, 8*) may be observed as they hover at honeysuckle flowers in the late evening. They are on the wing in June and July and there is normally only one brood in the year.

The eggs (*Fig 1*) are laid singly on leaves of the rosebay willowherb, or fireweed, (*Chamaenerion angustifolium*) and the great hairy willowherb (*Epilobium hirsutum*). The caterpillars hatch and are at first rather pale in colour, but soon they become dark and show the typical eyespots (*Fig 2*). These prominent eyespots and retractile front segments suggest an elephant's trunk; hence the common name of the species. When the larva is at rest the front segments are retracted, causing the segments with the eyespots to bulge and thus making the insect look much larger than it really is.

The larvae are nocturnal—they come up from their daytime retreat at dusk, and one may find them at this time by looking across the top of patches of the food plant. The pupa (*Fig 4*) is found in a loose cocoon (*Fig 3*) amongst the debris at the base of the food plant. The adult moths emerge in the following summer.

This species is found in central and southern Europe and through temperate Asia to Japan.

1 Eggs laid on leaf 2 Larva
3 Loose cocoon 4 Pupa
Plant:Rosebay willowherb
(Chamaenerion angustifolium)

94

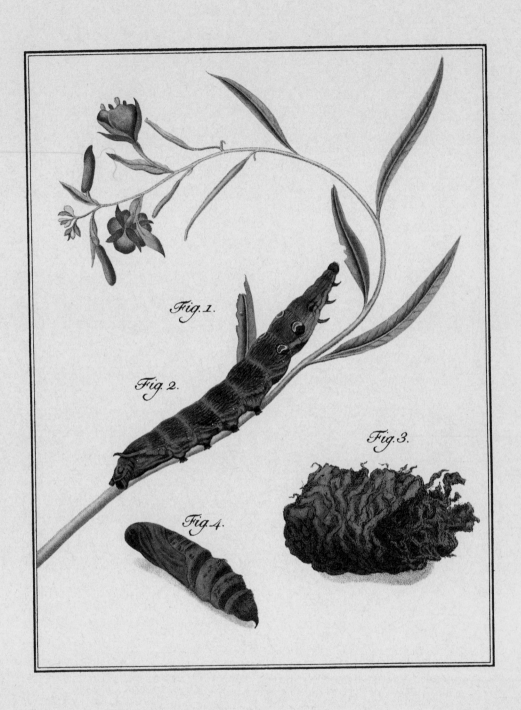

Fig. 1.

Fig. 2.

Fig. 3.

Fig. 4.

THE ELEPHANT HAWKMOTH

SPHINGIDAE *Deilephila elepnor* L.

(continued)

5 Pupal case after emergence of adult moth
6 Female moth 7 Female moth
8 Male moth

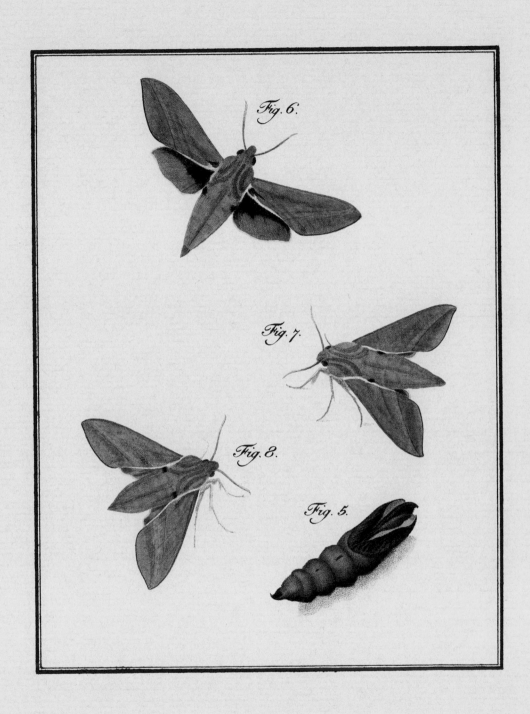

Fig. 6.

Fig. 7.

Fig. 8.

Fig. 5.

THE HUMMINGBIRD HAWKMOTH

SPHINGIDAE *Macroglossum stellatarum* L.

This striking moth is a day flying species in the main, although it is known to migrate at night as well as by day. The adult (*Figs 10, 11*) is often seen hovering in front of flowers in the garden, taking nectar by means of its long proboscis (*Fig 12*), a habit that gives it its common name. The Hummingbird Hawkmoth does not survive the winter here, but is a fairly regular migrant to our shores. Adults may arrive from southern Europe any time between June and October and probably breed in this country almost every year.

The eggs (*Figs 1, 2*) are laid on a variety of species of bedstraw, the preferred species being common yellow bedstraw (*Galium verum*). The larvae (*Figs 3, 4, 5, 6, 7, 8*) feed up very rapidly and soon spin a rudimentary cocoon in the litter under the food plant. Here they pupate (*Fig 9*). The long tongue is contained in a beak-like projection from the head of the pupa. The moth emerges from the pupa in warm weather about seven or eight weeks from the time when the eggs were laid.

This species is very widely distributed, being found throughout Europe, Asia as far as Japan and southern India and in North Africa. It may be seen anywhere in Britain, but is obviously most often met with in the south.

1 Egg, *natural size*
2 Egg *enlarged to show detail*
3-8 *Larvae* 9 *Pupa* 10-11 *Adult moths*
12 *Head of adult showing long tongue*
Plant: Goosegrass (Galium aparine)

Fig. 11.

Fig. 3.

Fig. 4.

Fig. 5.

Fig. 6.

Fig. 12.

Fig. 7.

Fig. 10.

Fig. 8.

Fig. 9.

Fig. 1.

Fig. 2.

THE PUSS MOTH

Notodontidae *Cerura vinula* L.

The Notodontidae, or 'Prominent' moths, have some of the most peculiar forms of caterpillar in the British Lepidoptera. The Puss Moth is a case in point. At rest, the larvae (*Figs 2, 3, 4, 5, 6, 7*) raise the fore and hind parts of the body and adopt a most odd pose. The fore part of the body has a hump or saddle, as do most species in this family, and the hind end bears a pair of whip-like flagellae, which are modified claspers. These flagellae can be withdrawn and erected rapidly, giving a rather snake-like appearance reminiscent of the osmaterium of the larvae of Swallowtail butterfly (*pages 26, 27*).

The larvae are easy to rear and may be found commonly on willow and sallow (*Salix* spp.) or poplar (*Populus* spp.) throughout Britain in July and August. When fully grown the caterpillar spins a very hard cocoon (*Fig 8*) on the bark of the tree on which it has fed. The cocoon has a thinner area at the head end and when the adult emerges from the pupa (*Fig 9*) in the following May or June it uses a ridge on the front part of the head to force its way out, helped by the secretion of potassium hydroxide from the mouth as a softening agent.

The adult moth (*Fig 11*) is fast flying and deposits its eggs (*Fig 1*) singly on the leaves of the food plant. It occurs widely abroad, being found from North Africa to Lapland and across Asia to Siberia and Japan.

1 Egg 2 Larva 3 Larva
4 Larva with tail filaments extended
5-7 Larvae 8 Cocoon 9 Pupa
10 Male moth 11 Female moth
Plant:Willow (Salix sp)

Fig. 10.

Fig. 11.

Fig. 4.

Fig. 3.

Fig. 5.

Fig. 2.

Fig. 6.

a

Fig. 7.

Fig. 1.

Fig. 9.

Fig. 8.

C. Sepp ad vir. del. et sculpsit.

THE LOBSTER MOTH

NOTODONTIDAE *Stauropus fagi* L.

The larvae (*Figs 1, 2, 3*) of this species, from which it gets its common name, are even more curious than those of the Puss moth. They rear the front and hind ends of the body almost at right angles to the central portion and wave their very long true legs and caudal flagellae in a most alarming way.

The adult moth (*Figs 6, 7, 8*) is on the wing in May and June and there is normally only one generation in the year. In very favourable seasons, however, a small second brood may be produced in August and September. The adult moth, unlike the Puss moth, may quite often be found sitting on the trunks of the trees in beech woods in southern England.

The eggs are laid singly on the leaves of the food plant and hatch after about ten days. The caterpillars, in common with many other species, eat the egg shells. Then they moult in a few days and begin to feed on the leaves of the food plant. They are normally found on beech (*Fagus sylvaticus*) but can also feed on birch (*Betula* spp.), oak (*Quercus* spp.) or hazel (*Corylus avellana*).

The caterpillars are fully grown by September and spin a cocoon between two dead leaves (*Fig 4*) where they moult to the pupal stage (*Fig 5*) and overwinter.

This interesting species is found in central and southern Europe and across Asia to Japan.

1 Larva 2 Larva 3 Larva
4 Cocoon spun between two leaves
Plant:Hazel (Corylus avellana)

Fig. 1.

Fig. 2.

Fig. 3.

Fig. 4.

THE LOBSTER MOTH
NOTODONTIDAE *Stauropus fagi* L.
(continued)

5 Pupa 6 Female moth 7 Male moth
8 Female moth
Plant:Hazel (Corylus avellana)

Fig. 8.

Fig. 7.

Fig. 5.

Fig. 6.

THE SWALLOW PROMINENT MOTH

NOTODONTIDAE *Pheosia tremula* L.

There are two species of Swallow Prominent, the Swallow Prominent (*Pheosia tremula*) and the Lesser Swallow Prominent (*P. gnoma* L.), the adults of which are extremely similar. The Swallow Prominent (*Figs 6, 7*) differs from the Lesser Swallow Prominent in that the last white wedge-shaped mark on the fore-wings is much narrower at the base.

There are two generations of both these species each year, and the adults are on the wing in May and June and again in August and September. The eggs (*Fig 1*) of the Swallow Prominent are laid singly on poplar (*Populus* spp.) or sallow *(Salix* spp.). The Lesser Swallow Prominent feeds on birch (*Betula* spp.)

The caterpillars (*Figs 2, 3*) feed up in about four or five weeks and in summer pupate (*Fig 5*) in a cocoon (*Fig 4*) near the surface, and at a greater depth in the autumn before overwintering. The caterpillar of the Lesser Swallow Prominent is brownish purple on top and violet beneath with a broad yellow stripe along the sides of the body.

Both species are found throughout much of the British Isles, but the Swallow Prominent is less common in the north while the Lesser Swallow Prominent is less common in the south. Both species occur through central and northern Europe and north-western Asia.

1 Egg on leaf 2 Larva 3 Larva
4 Cocoon 5 Pupa 6 Male moth
7 Female moth
Plant:Poplar (Populus sp)

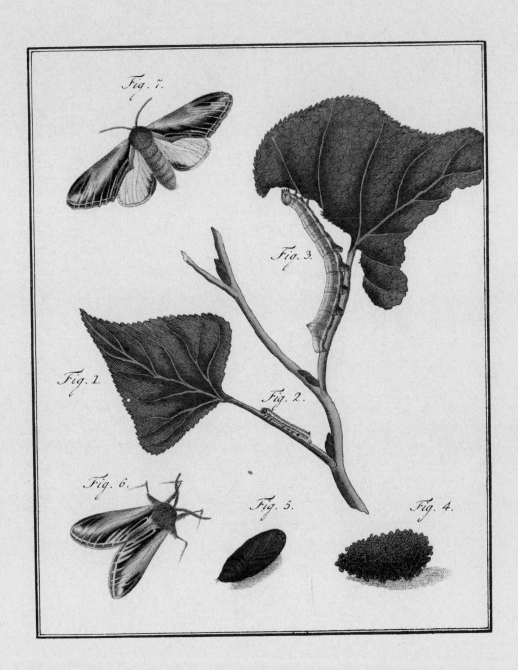

Fig. 7.

Fig. 3.

Fig. 1.

Fig. 2.

Fig. 6.

Fig. 5.

Fig. 4.

THE PEBBLE PROMINENT MOTH

NOTODONTIDAE *Notodonta ziczac* L.

The adults (*Figs 9, 10*) of this species are on the wing in May and June and again in late July and August. The vernacular name of this insect comes from the markings at the tips of the fore-wings which look like the outline of a pebble. The moth is found throughout Britain in damp and marshy habitats.

The eggs (*Fig 1*) are laid on willow and sallow (*Salix* spp.) and the curious caterpillars hatch in about a week. The shape of the caterpillars (*Figs 2, 3, 4, 5, 6, 7*) is somewhat like that of the Lobster moth but they lack the elongate fore-legs and long tail appendages. They have several sharp humps on the back. They feed up in four or five weeks and pupate in a cocoon (*Fig 8*) just under the soil surface.

Abroad, this species is found in central and northern Europe and as far south as Italy and eastwards to the Urals.

1 Egg 2-7 Larvae 8 Pupa in cocoon
9 Adult moth 10 Adult moth
Plant: Willow (Salix sp)

Fig. 10.

Fig. 6.

Fig. 5.

Fig. 4.

Fig. 2.

Fig. 3.

Fig. 1.

Fig. 9.

Fig. 7.

Fig. 8.

THE BUFF TIP MOTH

NOTODONTIDAE *Phalera bucephala* L.

The adults (*Figs 5, 6*) of this distinctive species are well camouflaged when at rest on the bark of a tree. The yellowish buff tips to the wings look very like the broken end of a branch. There is one generation per year which is on the wing in June and July.

The eggs (*Fig 1*) are laid in batches, often of thirty or forty, on the undersides of the leaves of a very large range of trees and shrubs. The caterpillars feed in compact groups (*Fig 2*) when small and do not roam very far even when fully grown (*Fig 3*). They become very easy to find because their combined feeding quickly results in the complete defoliation of a fairly large branch. Being hairy they are avoided by predatory birds and are also said to produce an unpleasant smell, though not all human beings can detect this.

The larvae become fully fed by late September and pupate (*Fig 4*) in the ground. They overwinter here and the adults emerge in the following June. The insect is common throughout the British Isles and may even be found in suburban gardens. Abroad, it occurs throughout Europe and Asia as far as Siberia and Asia Minor.

1 Eggs laid in a batch 2 Small larvae
3 Fully fed larva 4 Pupa 5 Adult moth
6 Adult moth
Plant: Willow (Salix sp)

Fig. 3.

Fig. 6.

Fig. 2.

Fig. 1.

Fig. 5.

Fig. 4.

THE PEACH BLOSSOM MOTH

Thyatiridae *Thyatira batis* L.

This very pretty species is on the wing from late May to late September. The adults (*Figs 7, 8, 9*) may be netted in the evening as they fly along the edges of woodland and hedgerows. There is one main generation in the year emerging in May, June and July, and a partial second brood in August and September.

The eggs are laid on the leaves of bramble and raspberry (*Rubus* spp.) and hatch in a few days. Some of the larvae (*Figs 1, 2, 3*) then feed up quickly to become fully fed in about four weeks, while others, even from the same batch of eggs, grow much more slowly and may take ten or twelve weeks to reach the same stage. The fully fed caterpillars then spin up in a leaf of the food plant (*Fig 4*) and moult to the pupal stage (*Fig 5*). The quick-growing larvae usually produce adult moths in August and September, while the slow-growing, as well as those from autumn adults, overwinter as pupae or occasionally even as well-grown larvae.

This species is common in wooded localities over much of Britain, but is definitely rarer in Scotland. It is found over most of northern and central Europe and Asia as far as Japan.

1 Larva 2 Larva 3 Larva
4 Pupa in cocoon in a curled leaf 5 Pupa
6 Detail of pupa (Cremaster)
7 Adult moth 8 Adult moth 9 Adult moth
Plant:Bramble (Rubus fruticosus)

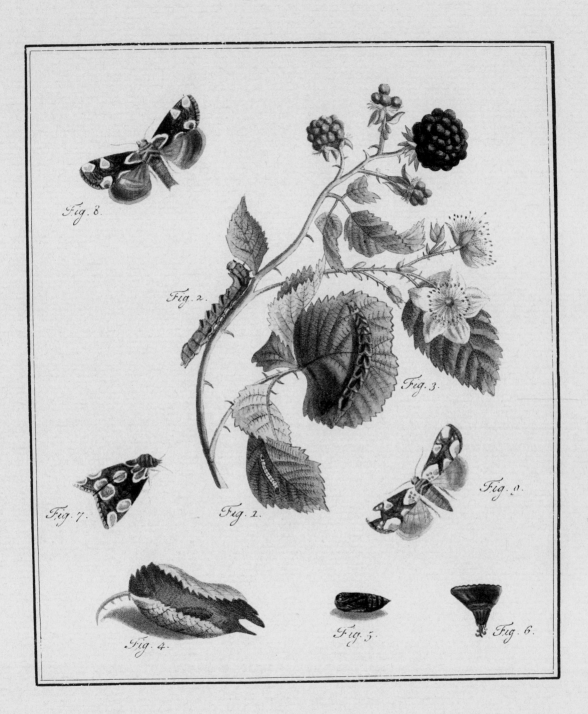

Fig. 8.

Fig. 2.

Fig. 3.

Fig. 7.

Fig. 1.

Fig. 9.

Fig. 4.

Fig. 5.

Fig. 6.

THE VAPOURER MOTH

LYMANTRIDAE *Orgyia antiqua* L.

This species is probably better known to town dwellers than to country dwellers since it often defoliates trees and shrubs in town gardens and streets.

The adults are about from June to October and are rather inconspicuous. The males (*Figs 10, 11*) are fully winged and fly to the females (*Fig 9*), which are probably the most degenerate of any British Lepidoptera. They are completely flightless and do not usually stray from the cocoon in which they pupated.

The eggs (*Figs 1, 2*) are normally deposited on the cocoon. The larvae hatch and, using a silken thread to float on wind currents, disperse to the food plant, which may be almost any tree or shrub.

The larvae (*Figs 3, 4, 5*) are very hairy and as with all species in this family, the hairs cause a nasty urticaria and should be handled with respect. They feed up quite quickly and are often pests in urban situations. They spin up on the bark of the food plant, fences, sheds or any other nearby rough surface. Hairs from the caterpillar are included in the cocoon (*Fig 6*) as a defence. The pupa (*Figs 7, 8*) may either hatch a week or two later, or overwinter, depending on the weather conditions. There are often several generations in the year.

This species is found throughout Britain—particularly in towns and by the sides of busy roads—in Europe and Asia as far east as Siberia and Tibet, and also in North America. It is a pest of ornamental trees and shrubs throughout most of its range.

1 Flightless female laying eggs
2 Egg enlarged to show detail
3-5 Larvae 6 Cocoon 7 Pupa, female
8 Pupa, male 9 Female moth
10 Male moth 11 Male moth
Plant:Willow (Salix sp)

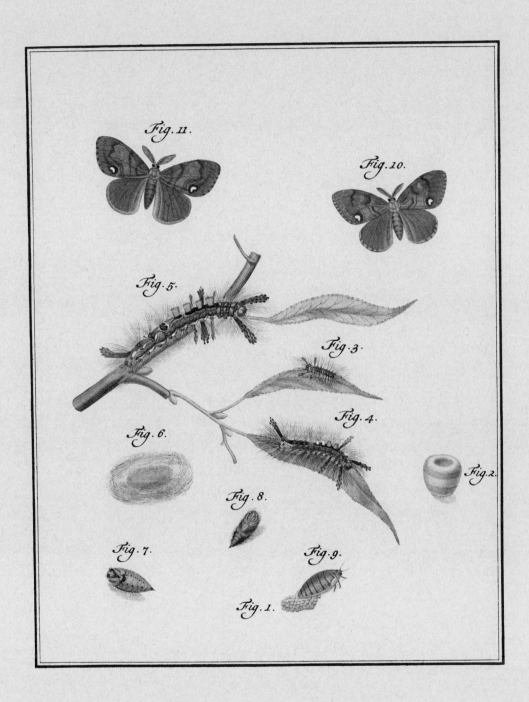

Fig. 11.

Fig. 10.

Fig. 5.

Fig. 3.

Fig. 4.

Fig. 2.

Fig. 6.

Fig. 8.

Fig. 7.

Fig. 9.

Fig. 1.

THE DECEMBER MOTH

LASIOCAMPIDAE *Poecilocampa populi* L.

This is another species which is probably commoner in urban situations than it is in the countryside. The adults (*Figs 6, 7*) are on the wing in October, November and December and may be seen fluttering around street lights near parks and gardens.

The eggs (*Figs 1, 2*) are laid on the bark of a wide range of trees and they do not hatch until the following spring. In April when the eggs hatch, the caterpillars (*Figs 3, 4*) begin to feed on the young foliage, completing their growth in June, before the foliage becomes too tough and chemically unsuitable. The pupae (*Fig 5*) are found in a loose cocoon either under loose bark or in crevices in the bark of the host trees or else in the leaf litter at their bases.

This species occurs throughout Britain, although it is more common in the south than in the north. It is found across northern and central Europe in woods, parks and suburban gardens.

1 Egg, natural size
2 Egg enlarged to show detail
3 Larva 4 Larva 5 Pupa 6 Male moth
7 Female moth
Plant:Oak (Quercus sp)

Fig. 6.

Fig. 7.

Fig. 3.

Fig. 4.

Fig. 1.

Fig. 2.

Fig. 5.

THE OAK EGGAR

LASIOCAMPIDAE *Lasiocampa quercus* L.

There are two sub-species of this moth, the Oak Eggar (*Lasiocampa quercus quercus*) in southern and lowland England and the Northern Eggar (*L. quercus callunae*) on the northern uplands from the Peak District and the Yorkshire Moors northwards. Both forms are sexually dimorphic, the males (*Fig 8*) being smaller and darker than the females (*Figs 9, 10*). The specimens illustrated show the Oak Eggar; the Northern Eggar is darker in colour and larger in size in both sexes. The adults of the southern form fly in July and August along hedgerows and the edges of woods, but the northern form is an inhabitant of moorlands and flies in May and June.

In the Oak Eggar the eggs (*Figs 1, 2*) are laid on a variety of trees and other plants, most commonly oak (*Quercus* spp.), willow (*Salix* spp.) and bramble (*Rubus fruticosus*). The caterpillars hatch, hibernate when quite small (*Figs 3, 4*) and then feed up in the following summer, to attain full size (*Fig 5*) by late June or early July. They then spin a large cocoon (*Fig 6*) in the litter under the food plant and the adults emerge from the pupa (*Fig 7*) in a few weeks.

The Northern Eggar on the other hand, feeds largely on heather (*Calluna vulgaris*) and has a two-year life cycle, spending the first winter as a small larva and the second as a pupa. The fully fed larvae of this form are often easy to find in the right habitat and, if reared, the virgin females of either form may be taken to the field and will attract males over quite long distances. This is because the females possess a scent, the sex pheromone, which attracts males to newly emerged females for mating. The males fly by day as well as night, whereas the females are only on the wing at night.

This species is widely distributed in Britain and is found throughout Europe.

1 Egg, natural size
2 Egg enlarged to show detail
3 Larva 4 Larva 5 Larva
6 Pupa in large cocoon
Plant:Willow (Salix sp)

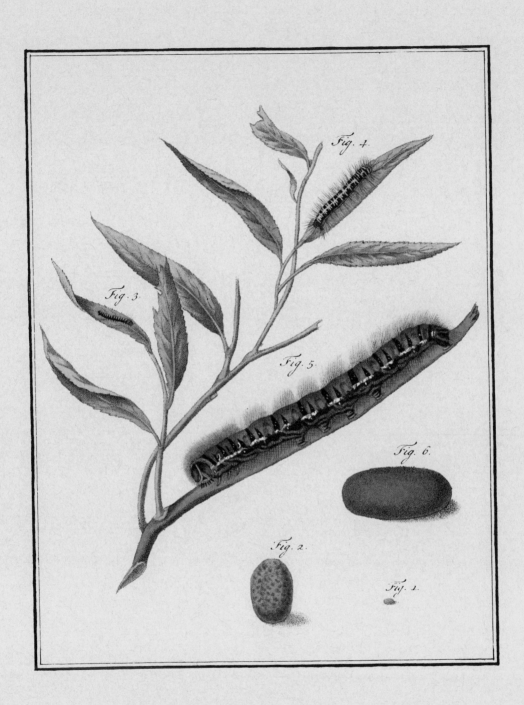

Fig. 4.

Fig. 3.

Fig. 5.

Fig. 6.

Fig. 2.

Fig. 1.

THE OAK EGGAR

LASIOCAMPIDAE *Lasiocampa quercus* L.

(continued)

*7 Pupa 8 Male moth 9 Female moth
10 Female moth*

Fig. 10.

Fig. 8.

Fig. 9.

Fig. 7.

THE FOX MOTH

LASIOCAMPIDAE *Macrothylacia rubi* L.

This handsome insect is, perhaps, our most familiar heath and moorland moth. The large adult moths (*Figs 11, 12*) are active both day and night and the fully grown larvae, which roll themselves into a ball when disturbed (*Fig 7*), are often to be found basking in the sun in early spring. These larvae should be handled with care as they have long urticating hairs which can cause painful weals on sensitive skin.

The eggs (*Fig 1*) are laid in batches on various plants, but perhaps most frequently on heather (*Calluna vulgaris*). They hatch in July and the caterpillar (*Figs 2, 3, 4, 5*) feeds up rapidly, becoming fully grown by October. It then hibernates for the winter and, although it does not feed again, it suns itself during the early part of the following spring. In April it spins a large cocoon (*Fig 9*) and pupates (*Fig 10*).

The adult moths emerge in May and June and the males can be seen flying low and fast over the moor in the afternoon searching for unmated females. The males (*Fig 11*) are smaller and darker than the females (*Fig 12*) and also have large feathery antennae. The females fly mainly at dusk and during the hours of darkness.

This species is found in suitable habitats throughout Britain and Europe.

1 Eggs 2-4 Larva
5 Larva rolled into a ball after being
disturbed
Plant: Willow (Salix sp)

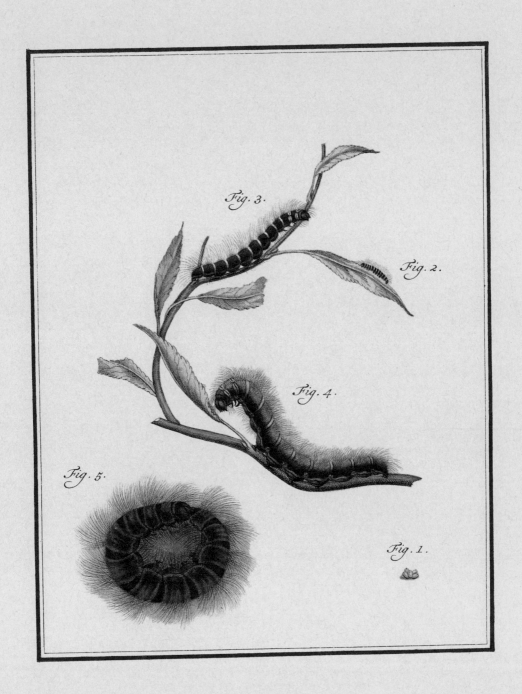

Fig. 3.

Fig. 2.

Fig. 4.

Fig. 5.

Fig. 1.

THE FOX MOTH

LASIOCAMPIDAE *Macrothylacria rubi* L.

(continued)

6 Larva 7 Larva rolled into a ball
8 Larva

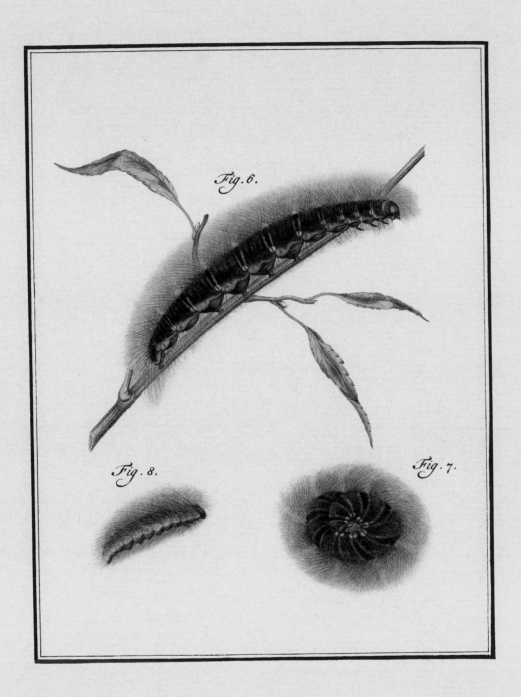

Fig. 6.

Fig. 8.

Fig. 7.

THE FOX MOTH

LASIOCAMPIDAE *Macrothylacria rubi* L.

(continued)

9 Large cocoon 10 Pupa 11 Male moth
12 Female moth
Plant:Willow (Salix sp)

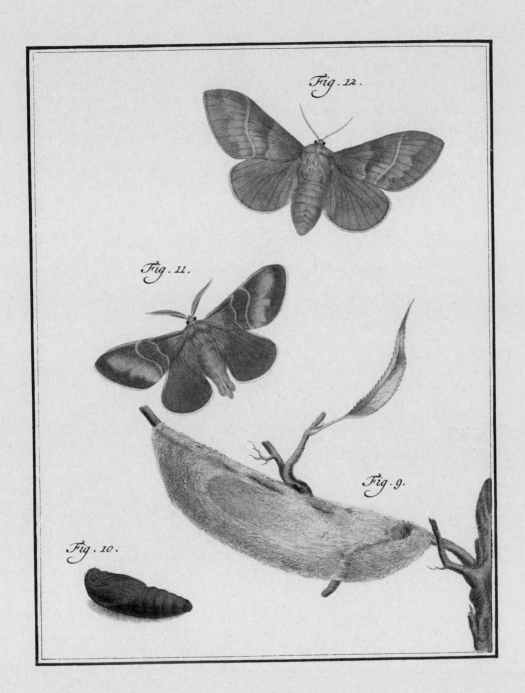

Fig. 12.

Fig. 11.

Fig. 9.

Fig. 10.

THE DRINKER MOTH

LASIOCAMPIDAE *Philudoria potatoria* L.

This species gets its English name and its scientific name from the larval habit of drinking dew and rain drops from the surface of its food plant.

The adult moths (*Figs 7, 8*) are on the wing in July and August and may be found in suitable damp habitats throughout Britain. The type of habitat does not seem to matter as long as it is damp, and this species may be found in meadows, moorlands and sand dunes wherever there is a good supply of the larger grasses, reeds and sedges. The adult males (*Fig 8*) are smaller and darker than the females (*Fig 7*) and have large feather-like antennae.

The eggs are laid in clusters (*Fig 1*) on a range of coarse grasses, reeds (*Phragmites communis*) and sedges (*Carex* spp.). The caterpillars (*Fig 3*) hatch in August and go into hibernation as partly grown larvae in October. They resume feeding in April and by early June they have completed their growth (*Fig 4*). They then spin a large cocoon (*Fig 5*) on the stem of a reed and pupate (*Fig 6*). The adults emerge after about four or five weeks.

This species is found in much of Europe and temperate Asia as far east as Japan.

1a, 1b Eggs laid in clusters 2-4 Larvae
5 Cocoon 6 Pupa 7 Female moth
8 Male moth
Plant:Reed (Phragmites communis)

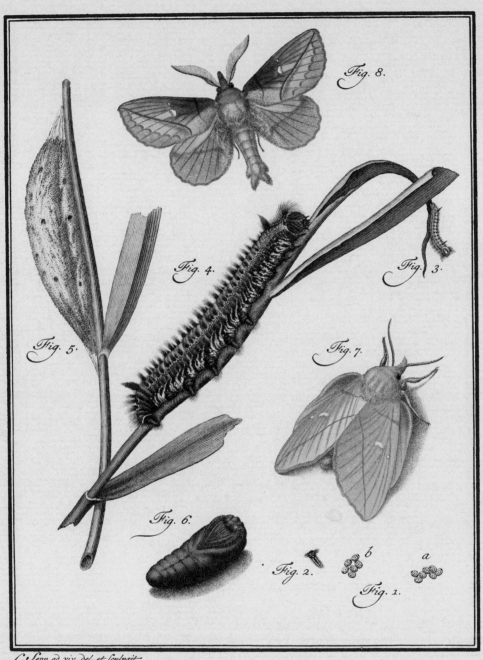

Fig. 8.

Fig. 4.

Fig. 3.

Fig. 5.

Fig. 7.

Fig. 6.

Fig. 2.

b

a

Fig. 1.

THE EMPEROR MOTH

SATURNIDAE *Saturnia pavonia* L.

This is our only representative of the Silk moth family, whose large and colourful foreign species fascinate many schoolboys.

The eggs (*Fig 1*) of this species are laid on a variety of plants, but perhaps the most frequent hosts are heather (*Calluna vulgaris*) and bramble (*Rubus fruticosus*). The caterpillars (*Figs 2, 3, 4, 5, 6, 7*) hatch in June and feed up quickly, becoming large, handsome fully fed larvae by mid-August. They then spin a large cocoon (*Fig 8*) in the food plant and pupate (*Fig 9*) for the winter.

The adults (*Figs 10, 11*) emerge in the following April and May. The males (*Fig 10*) have large antennae and are brightly coloured. They fly around over moors and heaths during the day with a fast low flight, and assemble in large numbers near newly emerged females, which emit a sex pheromone. The females (*Fig 11*) are larger than the males and less highly coloured; they fly mainly at night.

This species is found throughout the British Isles and Europe eastwards across Asia to Siberia.

1 Eggs 2-7 Larvae
Plant:Oak (Quercus sp) and Willow
(Salix sp)

130

Fig. 4.

Fig. 7.

Fig. 2.

Fig. 3.

Fig. 5.

Fig. 6.

Fig. 1.

THE EMPEROR MOTH

SATURNIDAE *Saturnia pavonia* L.

(continued)

8 Large cocoon spun in food plant
9 Pupa 10 Male moth 11 Female moth

Fig. 11.

Fig. 10.

Fig. 8.

Fig. 9.

THE PEBBLE HOOKTIP

DREPANIDAE *Drepana falcataria* L.

There are only six British species in this small and largely tropical family. All of them except one, the Chinese Character (*Cilix glaucata*) have the characteristic hooked tips to the wings which give the family its name, the Hooktips.

There are two generations a year of the Pebble Hooktip, the commonest member of this group. The adults (*Figs 7, 8*) are on the wing in May and June and again in August and September. They are found throughout most of Britain in areas of scrubby birch (*Betula* spp.) and alder trees (*Alnus glutinosa*). The caterpillars (*Figs 2, 3*), after hatching, roll the edges of the leaf on which they are feeding downwards. The edge is then held in position by a few strands of silk. They live in such shelters until they are nearly fully grown—some seven or eight weeks. They then spin a slight cocoon (*Fig 5*) in a rolled leaf and pupate (*Fig 6*). In the summer the adult moth emerges after a few days, while in those attaining full growth in September the pupal stage lasts over the winter until the following May.

This species occurs throughout Europe and into Asia.

1 Eggs 2 Larva 3 Larva
4 Detail of head of larva
5 Cocoon spun in rolled leaf 6 Pupa
7 Adult moth 8 Adult moth
Plant:Alder (Alnus glutinosa)

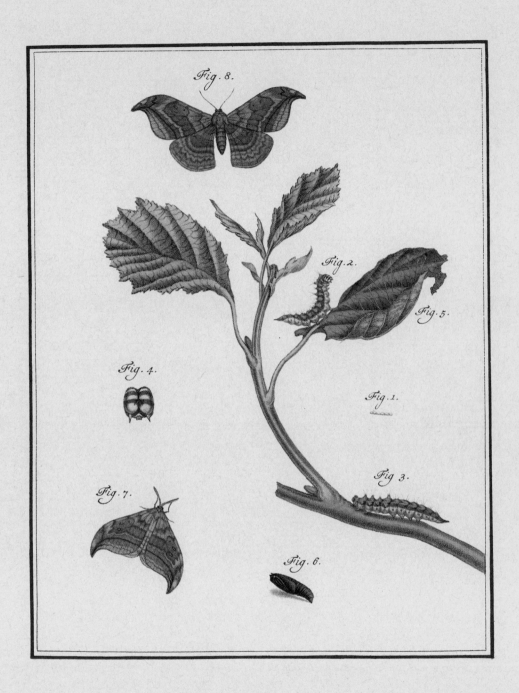

Fig. 8.

Fig. 2.

Fig. 5.

Fig. 4.

Fig. 1.

Fig 3.

Fig. 7.

Fig. 6.

THE GREEN SILVER-LINES MOTH

CYMBIDAE *Bena prasinana* L.

This pretty green species is common in oak, and mixed oak and beech, woodland throughout England and southern Scotland. It is often to be seen in those woods, sitting on ground layer vegetation such as bracken and other ferns.

The adults (*Figs 6, 7, 8*) are on the wing in June and July and lay their eggs (*Figs 1, 2*) on oak (*Quercus* spp.), birch (*Betula* spp.) or beech (*Fagus sylvaticus*). The caterpillars (*Fig 3*) feed up on the foliage and become fully fed by late September. They then spin a thin papery cocoon (*Fig 4*) on the undersurface of a leaf, or elsewhere on the tree, and there moult to the pupa (*Fig 5*). This pupa overwinters and the adult emerges in the following summer.

This species is widely distributed in suitable habitats across northern and central Europe and Asia to Japan.

1 Egg, natural size
2 Egg enlarged to show detail 3 Larva
4 Cocoon in leaf 5 Pupa 6 Adult moth
7 Adult moth 8 Adult moth
Plant:Oak (Quercus sp)

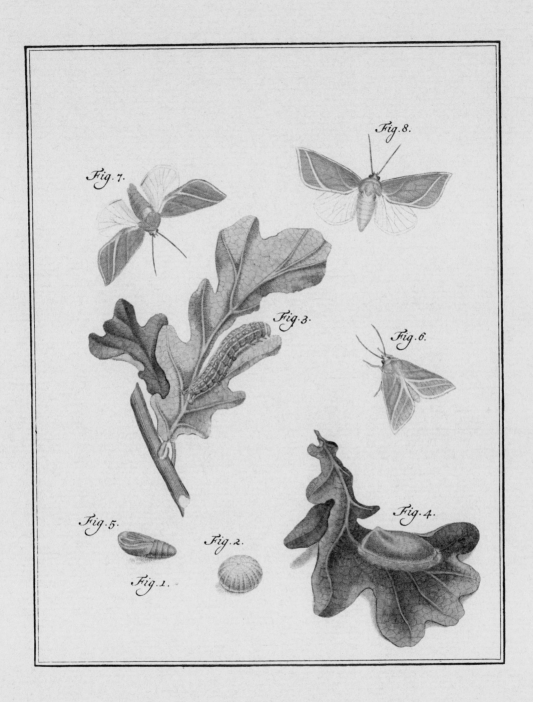

Fig. 8.

Fig. 7.

Fig. 3.

Fig. 6.

Fig. 5.

Fig. 2.

Fig. 1.

Fig. 4.

THE BUFF ERMINE MOTH

ARCTIIDAE *Spilosoma lutea* Hufn.

This moth is one of four similar species found in Britain, of which three, the White Ermine (*Spilosoma lubricipeda* L.), the Buff Ermine (*S. lutea*) and the Muslin moth (*Diaphora mendica* Cl.) are common throughout most of the country. The fourth species, the Water Ermine (*S. urticae* L.) is, its name suggests, more or less confined to the fens. The life history of all four species is very similar and, apart from the males of the Muslin moth, the adults are all white or buff in colour with varying degrees of spotting. The male Muslin moth is a dark smoky colour.

The Buff Ermine is on the wing in June and July and may be found sitting on fences or under street lamps. The males (*Fig 10*) are usually somewhat darker in colour than the females (*Fig 11*). The extent of the dark markings is very variable in both sexes and in some known varieties most of the wing area, apart from the veins, is a dark grey colour.

The eggs (*Figs 1, 2*) are laid on a wide range of cultivated and wild plants, but perhaps most frequently on nettle (*Urtica dioica*) and bramble (*Rubus fruticosus*). The larvae (*Figs 3, 4, 5, 6, 7*) hatch and feed up over the summer. They are the familiar "woolly bear" type of caterpillar characteristic of most species in this family. The fully grown larvae spin a rather loose cocoon (*Fig 8*) amongst leaf litter and overwinter as pupae (*Fig 9*).

This species is common in grassy places and gardens throughout the British Isles and its range extends across Europe and Asia to Korea and China.

1 Eggs, natural size
2 Eggs enlarged to show detail
3-7 Larvae 8 Cocoon in leaf litter
9 Pupa 10 Male moth 11 Female moth
Plant:Peaflower (Lathyrus sp)

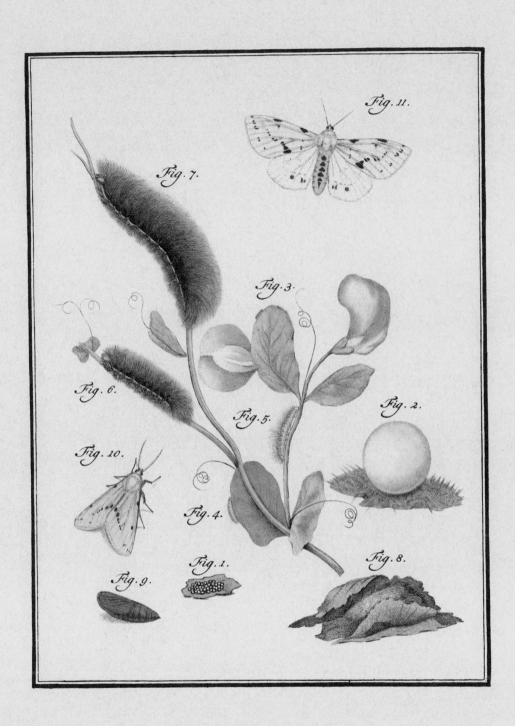

Fig. 11.

Fig. 7.

Fig. 3.

Fig. 6.

Fig. 5.

Fig. 2.

Fig. 10.

Fig. 4.

Fig. 8.

Fig. 1.

Fig. 9.

THE RUBY TIGER MOTH

ARCTIIDAE *Phragmatobia fuliginosa* L.

This species is common throughout Britain and may be found in a variety of habitats, but is perhaps most often met with on rough hill pastures and in marshy areas.

The adult moths (*Figs 7, 8*) are on the wing in May and June and again in August and September. Occasionally they can be seen flying during the daytime but they are basically nocturnal. The colouring varies considerably and one may observe a general tendency for the red colour to be replaced by a much darker brown or black as one moves northwards, so that specimens from northern Scotland have virtually no red on the wings and very little on the body.

The eggs (*Figs 1, 2*) are laid in small batches on the leaves of almost any low-growing plant, but most frequently on species of plantain (*Plantago* spp.) or dock (*Rumex* spp.). The caterpillars (*Figs 3, 4, 5*) feed up quickly and those from the spring generation of adults in the south pupate (*Fig 6*) within a brownish cocoon spun low down in the vegetation, and the adults emerge within a few weeks. In the north, and in the autumn generation in the south, the fully fed larvae overwinter and spin up in the early spring (April) without feeding any further; the adults emerge in May.

This species has a circumpolar distribution, being found across Europe, North Africa, northern and central Asia, Japan and North America.

1 Eggs laid on leaf
2 Egg enlarged to show detail
3-5 Larvae 6 Pupa 7 Adult moth
8 Adult moth
Plant:Cocksfoot (Dactylis glomerata)

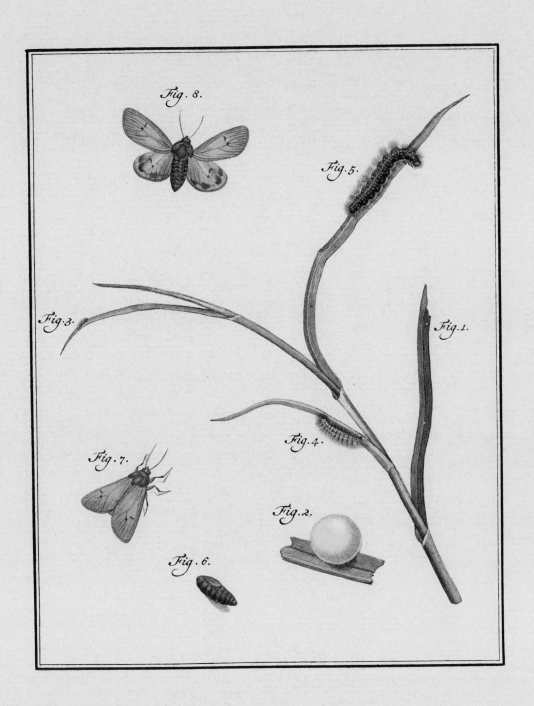

Fig. 8.

Fig. 5.

Fig. 3.

Fig. 1.

Fig. 7.

Fig. 4.

Fig. 2.

Fig. 6.

THE GARDEN TIGER MOTH

Arctiidae *Arctia caja* L.

The stage of this common and brightly coloured species most familiar to many people must be the caterpillar. It is the large "woolly bear" caterpillar (*Fig 3*) so often found in our gardens. It feeds on almost any low-growing wild or cultivated plant.

The eggs (*Fig 1*) are laid in batches on the leaves of the food plant in July and August. They quickly hatch and the small caterpillars (*Fig 2*) feed for about a month before overwintering. The caterpillars commence feeding again in April and become fully fed (*Fig 3*) in June. They then wander in search of a suitable place to spin their cocoon (*Fig 4*) and it is at this time that they often attract attention as they cross roads and pathways. The pupal stage (*Fig 5*) is fairly short and the adults (*Fig 6, 7*) emerge in July and August.

The large, brightly coloured adults are less often met with than might be expected from their size and gaudiness. They fly occasionally by day but usually at night. As with many other members of this group, they do not feed as adults and hence they do not visit flowers. They may be seen sitting on fences or under street lights to which they have been attracted. The females (*Fig 7*) are larger than the males (*Fig 6*). The extent of the dark markings on the wings is very variable, almost no two specimens being alike.

This species is very common in grassy places and in gardens even in cities and is found throughout Europe and temperate Asia as well as North America.

1 Eggs laid on leaf 2 Small larvae
3 Fully fed larva 4 Cocoon 5 Pupa
6 Male moth 7 Female moth
Plant:Nettle (Urtica dioica)

Fig. 7.

Fig. 6.

Fig. 3.

Fig. 2.

Fig. 4.

Fig. 5.

Fig. 1.

C. Sepp ad viv. del. et sculpsit.

THE CREAM-SPOT TIGER MOTH

ARCTIIDAE *Arctia villica* L.

This beautiful species is restricted to the southernmost counties of England and is perhaps more frequently seen near the southwest coast than elsewhere. It occurs along the edges of fields and on grassy waste-ground areas.

The adults (*Fig 6, 7*) fly in May and June and although it may occasionally be seen flying by day, it is normally active mainly at night. It is attracted to light, like many other night-flying Lepidoptera.

The eggs (*Fig 1*) are laid on a variety of low-growing herbs such as chickweed (*Cerastium* spp.), dock (*Rumex* spp.) and dandelion (*Taraxacum officinale*) and even sometimes on low gorse (*Ulex europaeus*) bushes. The caterpillars hatch and feed for a short time in July, after which they hibernate while still very small. They commence feeding again in the early spring and become fully fed (*Fig 2, 3*) by mid to late April. At this point they spin a loose cocoon (*Fig 4*) in the herbage and pupate (*Fig 5*). The adults emerge from the pupa three to four weeks later.

This species is found in much of western and southern Europe, usually most abundantly near the coast.

1 Egg, natural size 2 Larva 3 Larva
4 Loose cocoon 5 Pupa 6 Adult moth
7 Adult moth
Plant:Meadowgrass (Poa sp)

144

Fig. 6.

Fig. 2.

Fig. 7.

Fig. 3.

Fig. 4.

Fig. 1.

Fig. 5.

THE CINNABAR MOTH

ARCTIIDAE *Tyria jacobaeae* L.

This common day-flying species is familiar to many people both as the adult moth and as the larva. The larvae, with their black and yellow "football jersey" colouring are found wherever their food plants, the common ragwort (*Senecio jacobaea*) and the Oxford ragwort (*S. squalidus*) are found; even on odd plants growing at the edge of the pavement in the town centre.

The moth (*Figs 8, 9*) is on the wing in May and June and lays its eggs on either of the two common host plants. The caterpillars (*Figs 3, 4, 5, 6*) hatch and are very obvious, not only from their colouring, but also because they are frequently so numerous that they completely defoliate their food plant. Although so numerous, they are not attacked by vertebrate predators because of their noxious taste. This taste is due to chemicals acquired from their food; their bright yellow and black coloration is, in fact, a good example of "warning" coloration. The larvae feed up and become fully fed by August. They then pupate (*Fig 7*) in a rudimentary cocoon at or just below ground level. The adults emerge in the following spring and may be seen some distance from the food plant, so colonizing, at least in this country, even small patches of ragwort in isolated clumps.

This species is common throughout the British Isles and may be found wherever its food plant grows. Ragwort is a troublesome introduced weed of pastures in many parts of the world. In order to try to control it, Cinnabar moths have been introduced into a number of countries—e.g., Australia, New Zealand, South Africa and the United States—with mixed success. The natural range of the insect is throughout most of Europe and a large part of Asia.

1 Eggs, natural size
2 Egg enlarged to show detail
3-6 Larvae 7 Pupa 8 Adult moth
9 Adult moth
Plant:Groundsel (Senecio vulgaris)

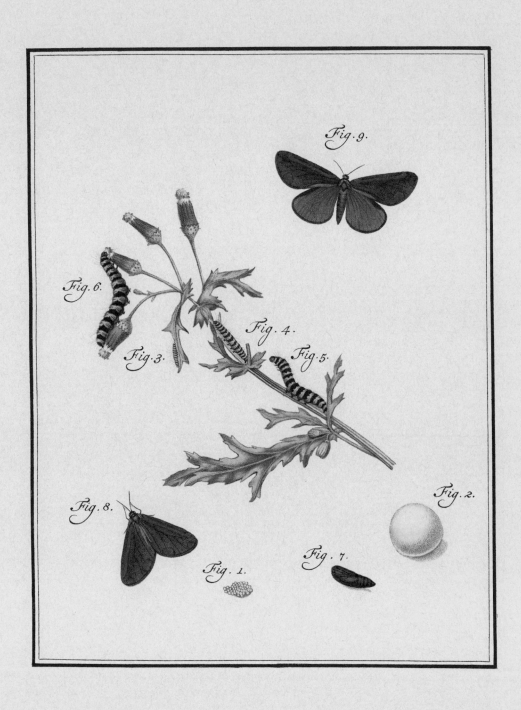

Fig. 9.

Fig. 6.

Fig. 4.

Fig. 3.

Fig. 5.

Fig. 2.

Fig. 8.

Fig. 7.

Fig. 1.

THE NUT-TREE TUSSOCK MOTH

NOCTUIDAE *Calocasia coryli* L.

This species is widely distributed throughout Britain and is locally common in rather scrubby exposed types of woodland and hedgerow. There are two generations in the year, the adults (*Figs 7, 8*) being on the wing in May and June and again in August and September.

The eggs (*Figs 1, 2, 3, 4*) are laid on a variety of trees such as oak (*Quercus* spp.), beech (*Fagus sylvatica*), and birch (*Betula* spp.), especially in exposed situations where these are stunted and bush-like in their growth.

The caterpillars (*Fig 5*) are variable in colour and may be anything from a dark brown to almost yellow. The tufts of long hairs may be greyish or red. The first brood feeds up in June and July and the second brood in September and October, before pupating (*Fig 6*) in a loose cocoon in leaf litter at the base of the food plant. The summer pupae emerge as adults within two or three weeks whereas the autumn pupae overwinter to emerge the following spring.

This species is distributed widely in Europe.

1 Egg 2 Egg
3-4 Eggs enlarged to show detail
5 Larva 6 Pupa 7 Adult moth
8 Adult moth
Plant:Oak (Quercus sp)

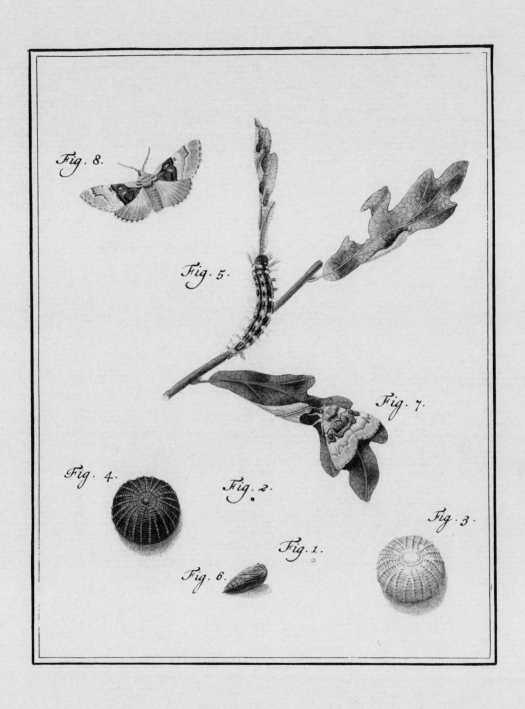

Fig. 8.

Fig. 5.

Fig. 7.

Fig. 4.

Fig. 2.

Fig. 3.

Fig. 1.

Fig. 6.

THE GREY DAGGER MOTH

NOCTUIDAE *Acronicta psi* L.

There are two species in Britain, the adults of which can only reliably be told apart by examining their genitalia—the Grey Dagger (*A. psi*) and the Dark Dagger (*A. tridens*). The larvae of these two species are, however, quite distinct and have different food preferences.

The adult moth (*Figs 5, 6*) is on the wing in May and June and often again in late August and September, at least in the southern part of the country. The eggs are laid on a variety of trees, particularly sallow (*Salix* spp.), poplar (*Populus* spp.), birch (*Betula* spp.) and apple (*Malus sylvestris*).

The caterpillars (*Figs 1, 2*) feed up rapidly and in four or five weeks they are ready to spin a cocoon (*Fig 3*) on the bark of the host tree in which to pupate (*Fig 4*). The summer pupae may emerge as adults in the autumn or overwinter to emerge in the following May.

This species is found throughout much of Europe and is widely distributed and common in Britain.

1 Larva 2 Larva
3 Cocoon spun on bark 4 Pupa
5 Adult moth 6 Adult moth
Plant:Oak (Quercus sp)

Fig. 6.

Fig. 1.

Fig. 2.

Fig. 5.

Fig. 3.

Fig. 4.

THE POWDERED WAINSCOT MOTH

NOCTUIDAE *Simyra albovenosa* Boeze.

The adults of this species (*Figs 8, 9*) are coloured so that, when at rest, they resemble dead reeds. Two unrelated groups of the Noctuidae look remarkably alike as adults, a phenomenon known as convergent evolution. The Powdered Wainscot represents one of the groups; the other much larger group includes several very common moths, for example the Common Wainscot (*Leucania pallens*) and the Smoky Wainscot (*L. impura*).

The Powdered Wainscot flies in June and is found, in this country, only in the fens of Norfolk and Cambridge. The eggs (*Figs 1, 2*) are laid in clusters on the leaves of various reeds and rushes, usually the common reed (*Phragmites communis*).

The hairy caterpillars (*Figs 3, 4, 5*) feed up on the leaves at night and are fully fed by September. They then spin a cocoon (*Fig 6*) on the food plant and the pupae (*Fig 7*) hatch in the following summer.

Abroad, this species is found in the marshes of northern and central Europe.

1 Eggs 2 Egg enlarged to show detail
3-5 Larvae 6 Cocoon 7 Pupa
8 Adult moth 9 Adult moth
Plant:Reed (Phragmites communis)

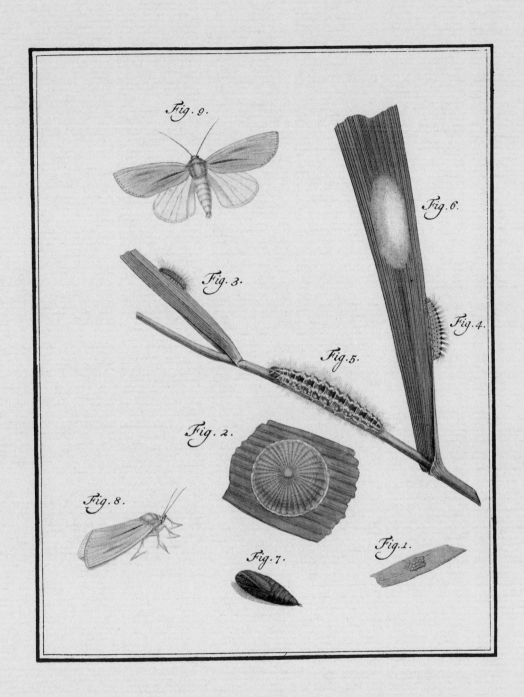

Fig. 9.

Fig. 6.

Fig. 3.

Fig. 4.

Fig. 5.

Fig. 2.

Fig. 8.

Fig. 7.

Fig. 1.

THE FLAME SHOULDER MOTH

NOCTUIDAE *Ochropleura plecta* L.

This species is common in a wide range of habitats throughout Britain. The adult moths (*Figs 6, 7*) may be met with at any time between April and September, but most frequently in May and June and again in August and September. There are at least two generations in the year and may often be at least a partial third. All three, however, overlap extensively and hence it is difficult to give limits to each.

The eggs (*Figs 1, 2*) are laid on any of a large number of low-growing broad-leaved plants, including at times some garden vegetables. The larvae (*Figs 3, 4*) feed up in three to four weeks and pupate (*Fig 5*) in the soil. The pupae may hatch within a few weeks or may lay over the winter, with the adults emerging in the following year.

This species is very widely distributed in Europe, Asia as far south as India, and North America.

1 Egg, natural size
2 Egg enlarged to show detail 3 Larva
4 Larva 5 Pupa 6 Adult moth
7 Adult moth
Plant: Dock (Rumex sp)

THE LARGE YELLOW UNDERWING MOTH

NOCTUIDAE *Noctua pronuba* L.

This species is, without doubt, one of the commonest of the larger moths in this country. Despite having only one generation in the year, it has a very long flight time. The adults may be found throughout the summer months, but there is a very definite peak of abundance in June and July.

The eggs are laid on almost any low-growing vegetation and the caterpillars (*Figs 1, 2, 3*) may overwinter in any stage from quite small larvae to fully fed individuals. They pupate (*Fig 4*) in the soil in the following spring. The caterpillars are very variable in colour. In the vegetable garden and occasionally in field crops, they can reach densities high enough for them to do considerable damage. Often, in this sort of situation, they act as "cutworms", cutting through the stems of plants by feeding at or just below soil level. They are only active at night, and hide by day in long grass or in any other dense vegetation.

The adults (*Figs 6, 7, 8, 9, 10*) are extremely variable in the colour of the fore-wings. They are often encountered when long grass is being cleared or when they are disturbed from the herbaceous border, and will then swiftly fly to a new daytime refuge.

This species is very common in many grassy habitats throughout Britain, perhaps especially around gardens and parks. It is found throughout Europe and North Africa and as far east as Siberia.

1-3 Larvae 4 Pupa
5 Detail of pupa (cremaster)
6 Adult moth
Plant:Pansy (Viola sp)

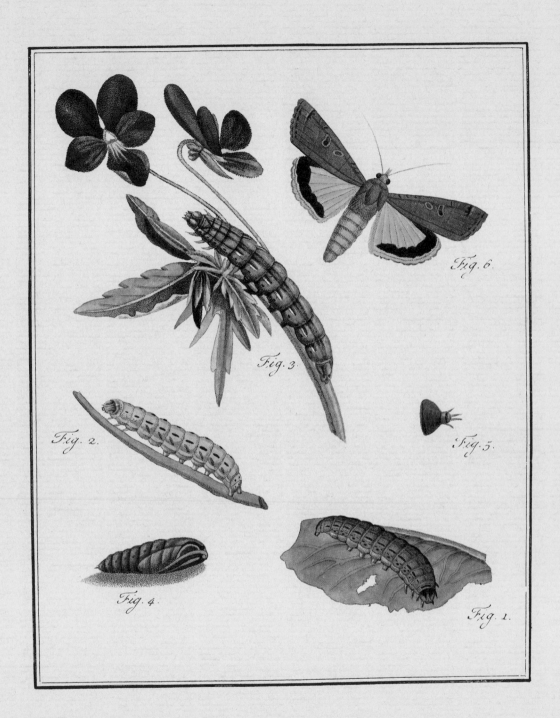

Fig. 6.

Fig. 3.

Fig. 2.

Fig. 5.

Fig. 4.

Fig. 1.

THE LARGE YELLOW UNDERWING MOTH

NOCTUIDAE *Noctua pronuba* L.

(continued)

7-10 Adult moths

Fig. 9.

Fig. 10.

Fig. 8.

Fig. 7.

THE DOT MOTH

NOCTUIDAE *Melanchra persicariae* L.

This species and the closely related Cabbage moth (*M. brassicae* L.) are often garden pests. The Cabbage moth is similar in general colour pattern to the Dot moth, but the main wing area is a lighter shade of grey and the light-coloured stigma is less well marked. It is far commoner than the Dot moth and is found throughout the British Isles feeding on a wide variety of plants, most often, perhaps, on Brassicas.

The Dot moth, though much less widespread, is common in much of England. The adults (*Figs 8, 9*) fly in July and August. The eggs are laid (*Figs 1, 2*) in batches on a very wide range of host plants. The caterpillars (*Figs 3, 4, 5, 6*) feed up in about two months and can often be found feeding on ornamental garden plants such as dahlias, chrysanthemums and anemones. The fully fed caterpillars pupate (*Fig 7*) in the soil.

This species, common in much of England though rare in the north, is found in a range of grassy habitats, but most frequently in suburban gardens. Abroad its distribution ranges across Europe to China and Japan.

1 Eggs laid on leaf
2 Egg enlarged to show detail
3-6 Larvae 7 Pupa 8 Adult moth
9 Adult moth
Plant:French marigold (Tagetes sp)

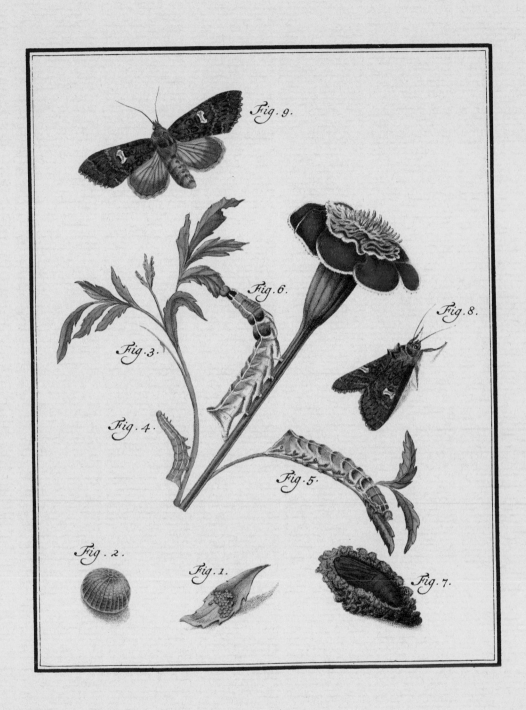

Fig. 9.

Fig. 6.

Fig. 8.

Fig. 3.

Fig. 4.

Fig. 5.

Fig. 2.

Fig. 1.

Fig. 7.

THE BRIGHT-LINE BROWN-EYE OR TOMATO MOTH

NOCTUIDAE *Diataraxia oleracea* L.

This is yet another pest species found throughout the British Isles. The adults (*Figs 6, 7*) fly in June and July and are rather variable as regards the ground colour of the fore-wings and the distinctness of the light-coloured line which gives the species its common English name.

The moth lays its eggs (*Figs 1, 2*) on a wide variety of plants, often on Brassica crops in gardens and fields. It is also a pest of tomatoes, not only defoliating the younger plants but also boring into the fruits. The caterpillars (*Figs 3, 4*) most often feed on foliage outside the greenhouse, but are also known to act as cutworms at times. They feed up in eight or nine weeks and pupate (*Fig 5*) in the soil where they overwinter.

This species is found throughout Europe and also in much of temperate Asia.

1 Eggs, natural size
2 Egg enlarged to show detail
3-4 Larvae 5 Pupa 6 Adult moth
7 Adult moth
Plant:Pellitory of the wall (Parietaria judaica)

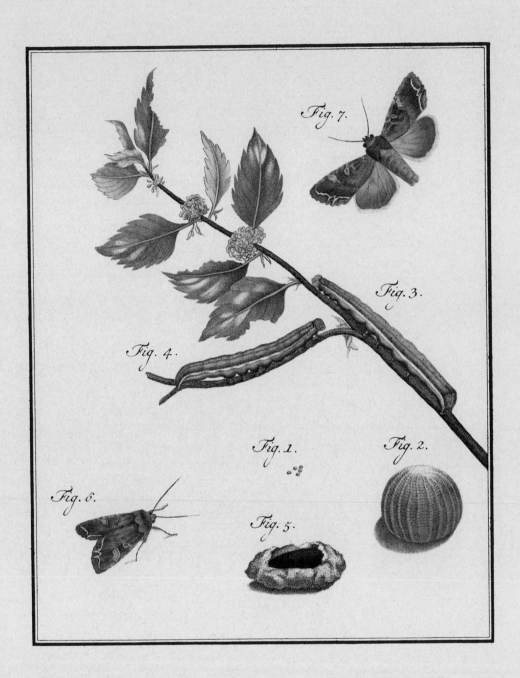

Fig. 7.

Fig. 3.

Fig. 4.

Fig. 1.

Fig. 2.

Fig. 6.

Fig. 5.

THE CAMPION MOTH

NOCTUIDAE *Hadena cucubali* Fuesl.

This is a common member of a group of very similarly coloured species, all of which feed as larvae on the seeds of various plants, usually species of campion or catchfly. The larvae often feed inside the unripe capsules and may be reared by collecting capsules which have frass exuding from them, indicating the presence of a caterpillar inside.

The adults (*Figs 5, 6*) are on the wing in June and July, and there is sometimes a partial second brood in August. They lay their eggs (*Figs 1, 2*) on a variety of species of campion, ragged robin and catchfly (*Silene* spp. and *Lychnis* spp.). The caterpillars (*Fig 3*) feed on the leaves and bore into the unripe seed pods, especially in the older stages, during July, August and September. They then pupate (*Fig 4*) in the soil and normally the adults emerge in the following year. In warm years, however, a proportion of the earlier pupae will emerge as adults in August.

This species and the very similar Lychnis moth (*H. bicruris*) are found almost throughout Britain and are common wherever suitable food plants grow. They occur through Europe and temperate Asia to Japan.

1 Egg, natural size
2 Egg enlarged to show detail 3 Larva
4 Pupa 5 Adult moth 6 Adult moth
Plant:Garden phlox

Fig. 5.

Fig. 6.

Fig. 1.

Fig. 3.

Fig. 2.

Fig. 4.

THE GREEN BRINDLED CRESCENT MOTH

NOCTUIDAE *Allophyes oxyacanthae* L.

The typical colour of this species is illustrated in *Figures 8* and *9*. The adults are on the wing in September and October and may be found commonly on the edges of woodland and in hedgerows and scrubland. A dark form of the adults has been known in this country for many years. Its wings are a uniform dark greyish brown, apart from the white crescent mark near the trailing edge of the fore-wings. This melanic form has recently become much more common in certain industrial areas where the darker coloration reduces predation on these individuals. This is because they are less visible on the dark bark of trees in polluted areas than the more typically coloured individuals. There are many other examples of this industrial melanism; e.g., the Lobster moth (*pages 102-105*), the Oak Eggar (*pages 118-121*) and, most famous of all, the Peppered moth (*pages 226-229*).

The eggs (*Figs 1, 2, 3*) are laid on the trunk and twigs of the host plant: hawthorn (*Crataegus oxyacanthoides*), sloe (*Prunus spinosa*), and apple (*Malus sylvestris*). They do not hatch until the following spring and the caterpillars (*Figs 4, 5, 6*), of which a black melanic form is known, feed up rapidly, becoming fully fed by late May or early June. They then pupate (*Fig 7*) in the soil at the base of the tree on which they have been living.

This species is common throughout most of Britain and the dark form is the commonest form in some urban areas, London for example, but also turns up frequently in some rural areas. Although this dark form is unknown outside Britain, the species is found throughout northern and central Europe eastwards to Siberia.

1 Egg, natural size
2-3 Eggs enlarged to show detail
4-6 Larvae 7 Pupa 8 Adult moth
9 Adult moth
Plant:Hawthorn (Crataegus sp)

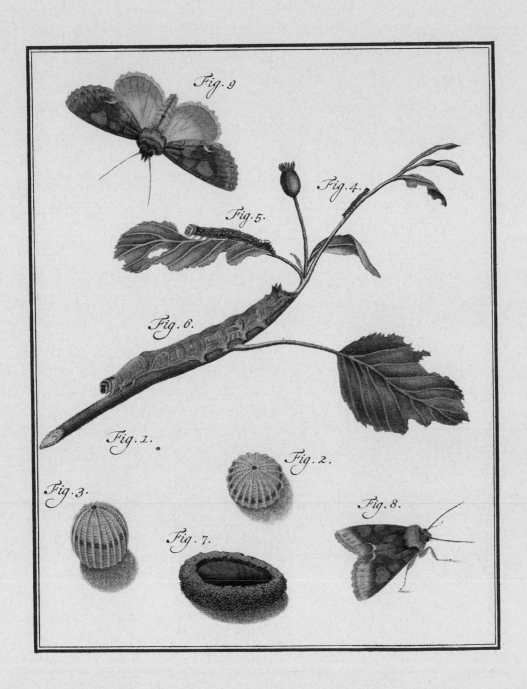

Fig. 9.

Fig. 4.

Fig. 5.

Fig. 6.

Fig. 1.

Fig. 2.

Fig. 3.

Fig. 7.

Fig. 8.

THE MERVEILLE DU JOUR MOTH

NOCTUIDAE *Griposia aprilina* L.

Both the caterpillars (*Fig 6*) and the adults (*Figs 10, 11*) of this pretty species are difficult to detect when they are at rest during the day on the trunks and branches of lichened oak trees. The adult moths are on the wing in September and October and may be found in oak woods throughout Britain.

The eggs are laid on twigs of the host tree, oak (*Quercus* spp.), and overwinter to hatch in the following spring. The larvae hatch in April and feed until late May or early June, when they descend from the tree to pupate (*Fig 9*) in a cocoon (*Fig 8*) under moss at the base of the tree.

The moth is common in oak woods throughout Britain and Europe.

6 Larva 7 Detail of head of larva
8 Cocoon 9 Pupa 10 Adult moth
11 Adult moth
Plant: Lichened bark

Fig. 11.

Fig. 6.

Fig. 7.

Fig. 8.

Fig. 9.

Fig. 10.

THE ANGLE-SHADES MOTH

NOCTUIDAE *Phlogophora meticulosa* L.

The adults (*Figs 6, 7*) of this very common species may be found at any time of the year, but most commonly in May and June and again in September and October. When they are newly emerged from the pupa they have beautiful shades of pink and green on the wings, but these fade rapidly as the moth ages and after death. At rest during the daytime the moth sits with its wings crumpled in a longitudinal plane and looks very like a dead leaf.

The eggs (*Figs 1, 2*) are laid on a wide range of plants—even on such things as geraniums in the greenhouse—but perhaps most commonly on low plants such as deadnettle (*Lamium* spp.), dock (*Rumex* spp.) and groundsel (*Senecio vulgaris*). The larvae (*Figs 3, 4*) feed up quickly in spring and summer, and pupate (*Fig 5*) in leaf litter. This species can overwinter as a larva, pupa or adult and hence, although the double peak in numbers of adults suggests that there are normally two generations in the year, these overlap greatly. The autumn generation of moths over-winters as adults or larvae but the majority of specimens overwinter as the pupa.

This moth is also known as a migrant, but how far this affects the numbers in this country is not known. The species is found commonly throughout Britain and much of Europe, North Africa and parts of the Middle East.

1 Egg, natural size
2 Egg enlarged to show detail
3-4 Larvae 5 Pupa 6-7 Adult moths
Plant:Dead nettle (Lamium sp)

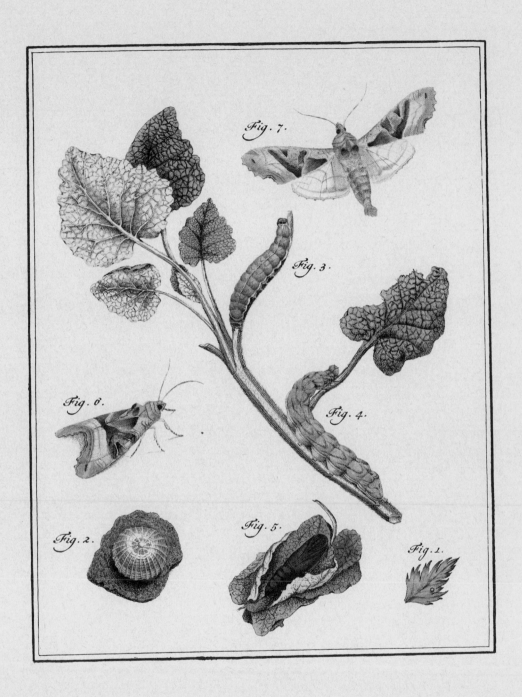

Fig. 7.

Fig. 3.

Fig. 6.

Fig. 4.

Fig. 2.

Fig. 5.

Fig. 1.

THE COPPER UNDERWING MOTH

NOCTUIDAE *Amphipyra pyramidea* L.

This species, although common in the southern part of the country, is absent or rare in the north of England and in Scotland. The adults (*Figs 3, 5*) are on the wing in August and September. The eggs are laid on the twigs of a range of trees and shrubs, but most commonly on oak (*Quercus* spp.), birch (*Betula* spp.) or sallow (*Salix* spp.). The larvae (*Fig 1*) hatch in the following April and are fully grown by early June. The pupa (*Fig 2*) is in a light cocoon inside a rolled leaf.

The fly shown in *Figure 4* is a tachinid fly and is a parasite of the larvae. This type of fly is commonly found as a parasite of Lepidoptera and other insects.

The adult moths of the Copper Underwing roost in groups during the day behind pieces of loose bark on trees. The species is found across much of central and southern Europe and in North America.

1 Larva 2 Cocoon and pupa
3 Adult moth 4 Parasitic tachnid fly
5 Adult moth
Plant:Oak (Quercus sp)

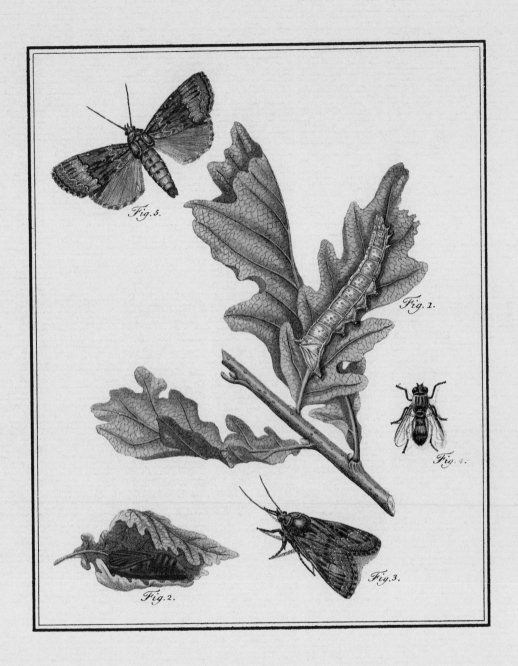

Fig. 5.

Fig. 1.

Fig. 4.

Fig. 2.

Fig. 3.

THE MOUSE MOTH

Noctuidae *Amphipyra tragopogonis* L.

This rather sombre-coloured moth is named, not only for its colour, but also because of its habit, when disturbed in the daytime, of scuttling along the ground rather than flying to find another place to rest. It is very common on the edges of woods and in gardens and waste ground throughout Britain.

The eggs are laid in small batches on many plants, both woody and herbaceous plants; for example sallow (*Salix* spp.), hawthorn (*Crataegus oxyacanthoides*), plantain (*Plantago lanceolata*) and many ornamental garden plants.

The caterpillars (*Fig 1*) hatch in the following spring and feed from April until early June. They then spin up (*Fig 2*) and pupate (*Fig 3*), and the adults emerge in some three to four weeks.

This species is common throughout Britain and is widely distributed abroad, being found in most of Europe, parts of Asia and in the eastern United States.

1 Larva 2 Cocoon 3 Pupa
4 Detail of pupa (cremaster)
5 Adult moth 6 Adult moth
Plant:Larkspur (Delphinium ambiguum)

Fig 6

Fig. 5.

Fig. 1.

Fig 4.

Fig. 3.

Fig 2.

THE HEBREW CHARACTER MOTH

NOCTUIDAE *Orthosia gothica* L.

This species and the Common Quaker moth that follows are the commonest of the noctuid moths that fly in the early spring. These and several other species may be taken by going at night with a torch to sallow bushes when they have mature catkins ("pussy willows"). Since there are few other sources of nectar and pollen at that time of year many insects congregate around these bushes to feed.

The adult moths (*Figs 6, 7*) are on the wing from March to early May. They lay their eggs (*Figs 1, 2*) in batches, often two deep, on the branches of a number of low plants, trees and bushes, but most commonly on oak (*Quercus* spp.), sallow and willow (*Salix* spp.). The caterpillars (*Figs 3, 4, 5*) hatch in late April or early May and feed until late June or July. The fully fed caterpillars then descend from the trees and pupate (*Fig 6*) in the soil. The pupae overwinter and the adults emerge in the following spring.

This species is very common over the whole of the British Isles, and through Europe and temperate Asia to Japan.

1 Egg, natural size
2 Egg enlarged to show detail
3-5 Larvae 6 Pupa 7-8 Adult moths
Plant: Willow (Salix sp)

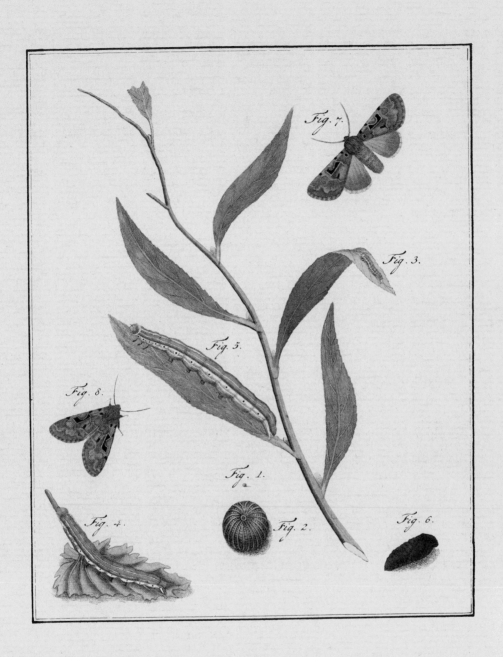

THE COMMON QUAKER MOTH

NOCTUIDAE *Orthosia stabilis* View.

This species, as common as, if not commoner than the last species, can be found as an adult (*Figs 6, 7*) feeding at sallow catkins during March and April. It is very variable both in the ground colour of the wings and in the degree of dark markings on the fore-wings. The general colour may vary from pale buff to dark brown. This variability makes it difficult to distinguish from several other species, some of which are very common, on the wing at the same time of year.

The eggs (*Figs 1, 2*) are laid in small groups on the twigs of various trees, but again oak (*Quercus* spp.), sallow (*Salix* spp.) and birch (*Betula* spp.) seem to be preferred. They hatch in April and the caterpillars (*Fig 3*) feed until late May or early June. They then descend the tree and pupate (*Fig 5*) in a cocoon (*Fig 4*) beneath moss or in the ground. The moths emerge in the following spring.

This species is very common throughout Britain and its range abroad extends across southern and central Europe to Asia Minor.

1 Eggs, natural size
2 Egg enlarged to show detail 3 Larva
4 Cocoon 5 Pupa 6-7 Adult moths
Plant:Oak (Quercus sp)

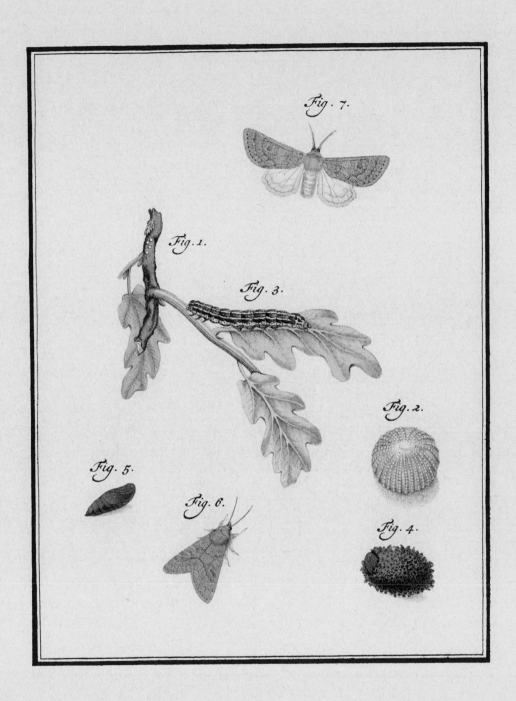

Fig. 7.

Fig. 1.

Fig. 3.

Fig. 2.

Fig. 5.

Fig. 6.

Fig. 4.

THE DUNBAR MOTH

NOCTUIDAE *Cosmia trapezina* L.

This very common species is on the wing in July and August. The adults (*Figs 5, 6, 7*) may be found around lights in any area of woodland.

The eggs are laid in small batches on the twigs of a wide range of trees, especially oak (*Quercus* spp.), sallow (*Salix* spp.) and elm (*Ulmus* spp.), and the caterpillar hatches from the overwintered eggs in the following spring. The caterpillar (*Figs 1, 2*) then feeds, not only on the foliage of the host tree, but also on the caterpillars of other moths, the Winter moth (*pp 204, 205*) in particular. This makes it one of the very few regularly carnivorous Lepidoptera in this country. The caterpillars are fully fed by early June and descend the tree to pupate (*Figs 3, 4*) in the soil. The moths emerge after about a month.

In Britain this species is common almost wherever there are suitable trees, and abroad it is found across Europe and temperate Asia to Japan.

1-2 Larvae 3-4 Pupae 5-7 Adult moths
Plant:Elm (Ulmus sp)

Fig. 5.

Fig. 7.

Fig. 1.

Fig. 2.

Fig. 3.

Fig. 4.

Fig. 6.

THE RED SWORDGRASS MOTH

NOCTUIDAE *Xylena vetusta* Hbn.

There are two species in this group, the Red Swordgrass (*X. vetusta*) illustrated in *Figures 3* and *4* and the Swordgrass moth (*X. exsoleta*). The adults of both species are similar in basic pattern, but the Red Swordgrass has much more of the dark coloration on the hind area of the forewings, and the black wedges near the wing edge are much larger.

The adults of both species are on the wing in the autumn from late September until October or even November, when they go into hibernation. They may be seen again at sallow blossom in March and April.

The eggs of both species are laid on a wide range of herbaceous plants, but the Red Swordgrass is commonest on rushes (*Cyperaceae*) and on yellow flag (*Iris pseudacorus*). The caterpillars (*Fig 1*) are brightly marked and feed up in eight to ten weeks. They then pupate (*Fig 2*) in the soil in July and the adult emerges some two months later.

Both species in this group are widespread in their distribution, being found almost throughout Britain. Neither is very common anywhere but both are more common in the north than in the south of the country. They are found in damp woods, and on marshes and waste ground. Abroad, they range throughout much of Europe and Asia, and the Red Swordgrass also occurs in North America.

1 Larva 2 Pupa 3-4 Adult moths
Plant:Water dropwort (Oenanthe fluviatilis)

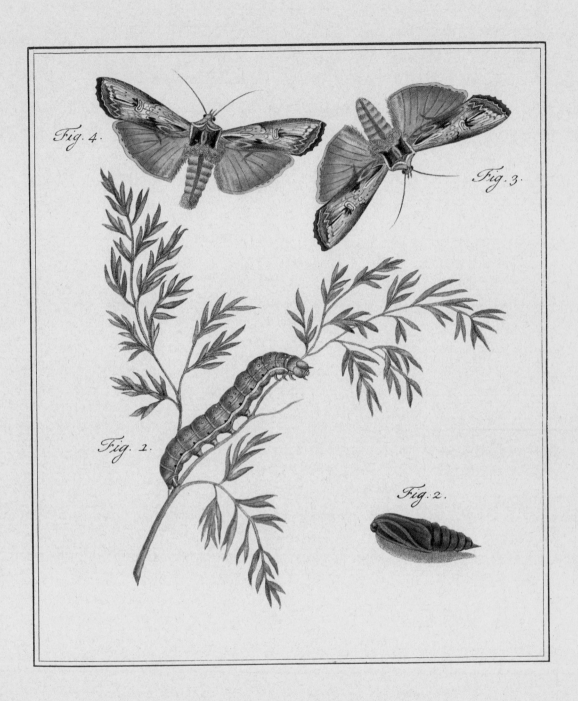

Fig. 4.

Fig. 3.

Fig. 1.

Fig. 2.

THE BEAUTIFUL YELLOW UNDERWING MOTH

NOCTUIDAE *Anartia myrtilli* L.

This small, fast-flying species is extremely difficult to follow as it flies in the heat of the day across heather moors and heaths. The colour of the fore-wings blends perfectly with the heather and the yellow flash of the hind-wings serves to divert the eyes as it takes off and lands.

The adults (*Figs 3, 4, 5*) may be seen at any time in the summer but are most common in June and July. The eggs are laid on heather (*Calluna vulgaris*) or heath (*Erica* spp.). The caterpillars (*Fig 1*) hatch and feed up on the food plant during the summer. Occasionally they may overwinter and complete their development in the spring, but usually the fully fed larvae pupate (*Fig 2*) in September and October and overwinter as pupae.

This species is common on heaths and moors throughout Britain and in northern and central Europe and parts of Atlantic North America.

1 Larva 2 Pupa 3-5 Adult moths
Plant:Heather (Calluna vulgaris)

Fig. 4.

Fig. 5.

Fig. 1.

Fig. 3.

Fig. 2.

THE BORDERED SALLOW MOTH

NOCTUIDAE *Pyrrhia umbra* Hufn.

This species has an extremely wide range; being circumpolar in its distribution, it occurs in temperate Europe, Asia and North America. In Britain it is found as far north as the Great Glen, but is nowhere very common, though it is said to be more plentiful near the coast than inland.

The adults (*Figs 5, 6*) are on the wing in May and June and there is only one generation in the year. It may be found feeding on flowers, particularly low-growing plants, the preferred hosts being restharrow (*Ononis repens*), a plant of sandy areas, and also various species of *Geranium*, including the cultivated varieties. The caterpillars (*Figs 1, 2*) feed during July and August and pupate (*Fig 3*) in the ground, where they overwinter. The adult moth emerges the following summer.

1 Larva 2 Larva 3 Pupa
4 Cremaster (pupal hook-like attachment)
5-6 Adult moths
Plant:Pelargonium

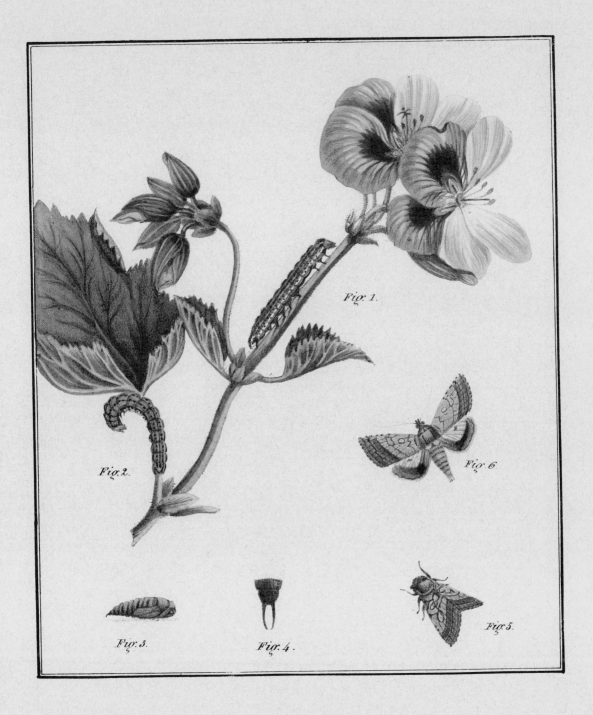

Fig. 1.

Fig. 2.

Fig. 6.

Fig. 3.

Fig. 4.

Fig. 5.

THE HERALD MOTH

NOCTUIDAE *Scoliopteryx libatrix* L.

This species too has a circumpolar distribution. It is common throughout Britain.

The adults (*Figs 4, 5*) may be found on the wing in August, September and October; these then hibernate in hollow trees, sheds, etc. to reappear early in the spring. They are often amongst the first moths to emerge from hibernation, hence the English name of the species. The adults are on the wing right through until June. They may be seen at dusk feeding at ivy blossom or piercing the fruits of bramble, or at sallow in the spring. They may also be attracted to "sugaring mixture" (a mixture of treacle, fruit juice, honey and stale beer) painted on fence posts and tree trunks. This mixture will attract many species of moths and other insects to feed at those times of year when there are few other sources of nectar available in nature.

There is probably at least a partial brood from the early eggs laid at the beginning of April, producing new adults in late May and early June. The main part of the population is, however, single-brooded, the eggs (*Fig 1*) being laid in April and May on sallows (*Salix* spp.) and poplars (*Populus* spp.). The caterpillars (*Fig 2*) feed up to become fully grown after two to three months. They then spin two leaves together (*Fig 3*) and pupate. The adults emerge a few weeks later.

1 Egg, natural size 2 Larva 3 Pupa
4-5 Adult moths
Plant: Willow (Salix sp)

188

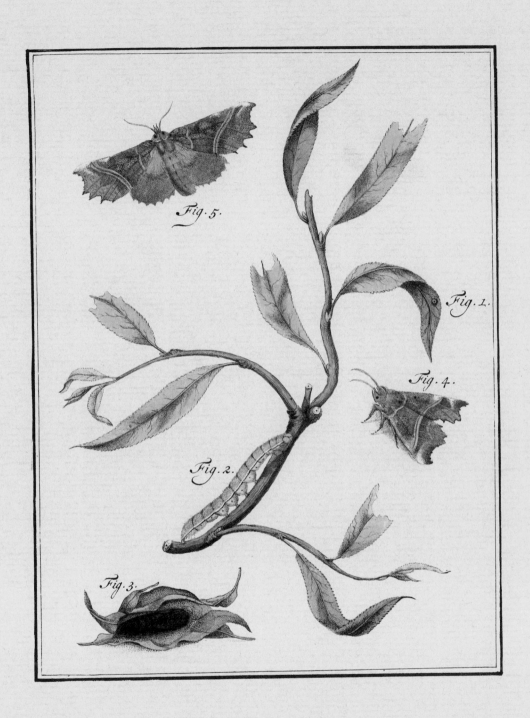

Fig. 5.

Fig. 1.

Fig. 4.

Fig. 2.

Fig. 3.

THE BURNISHED BRASS MOTH & THE SILVER Y MOTH

Noctuidae *Plusia chrysitis* L. & *Plusia gamma* L.

These two pretty species are the commonest of a group of moths that have metallic silver or gold marks on the wings. The adults of the well-named Burnished Brass (*Figs 11, 12*) are on the wing in June, July, August and September and may be seen in the evening visiting garden and hedgerow flowers. There are two overlapping generations in the year. The first emerges in July and lays its eggs (*Figs 7, 8*) on many low-growing plants, e.g., nettle (*Urtica dioica*), deadnettle (*Lamium* spp.), etc.; the second generation emerges in July and August. The caterpillars (*Fig 9*) of the first generation feed up rapidly and pupate (*Fig 10*) in a cocoon amongst the food plant within about four weeks, while the autumn generation feed more slowly and overwinter as partly grown larvae to pupate the following May. This species is common along hedges and on waste ground throughout Britain and in most of Europe.

The Silver Y (*Figs 5, 6*) is a very common day-flying species that can be seen visiting flowers in almost any garden on sunny days from May to September. It does not survive the winter here but migrates to our shores from southern Europe and North Africa in large numbers almost every year. The spring migrants breed here, sometimes in large enough numbers to be a pest of food and ornamental plants. The adults are always much more numerous in the autumn than in the spring, although how much this is due to breeding in this country and how much to further immigration is not clear.

The eggs (*Figs 1, 2*) are laid on a wide variety of wild and cultivated plants. The caterpillars (*Fig 3*) have two forms, a light form as in the plate and a dark form which occurs when the density of larvae is very high. This type of colour change associated with increased density of animals and increased migratory activity of the subsequent adults is an example of phase change, analogous to that of the locust. The pupa (*Fig 4*) is found in a cocoon spun into the leaves of the host plant.

1 Egg, natural size 2 Egg enlarged
3 Larva 4 Pupa 5-6 Adult moths
7 Egg, natural size 8 Egg enlarged
9 Larva 10 Pupa 11-12 Adult moths

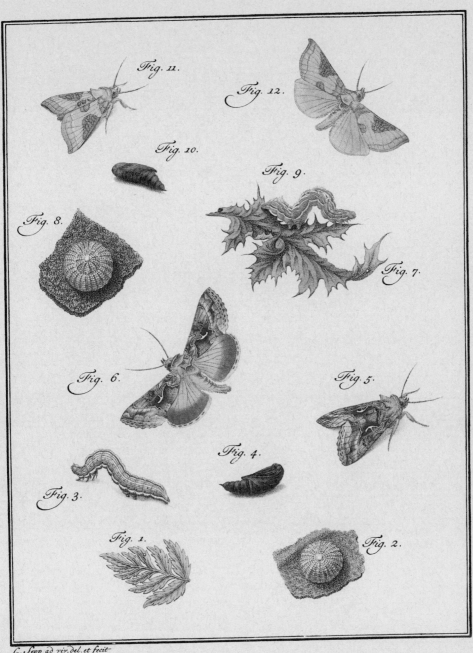

Fig. 11.

Fig. 12.

Fig. 10.

Fig. 9.

Fig. 8.

Fig. 7.

Fig. 6.

Fig. 5.

Fig. 4.

Fig. 3.

Fig. 1.

Fig. 2.

C. Sepp ad viv. del. et fecit.

THE MOTHER SHIPTON MOTH

Noctuidae *Euclidimera mi* Cl.

There appears to be only one generation of this fast day-flying moth in this country, although on the continent, in Holland for example, there are normally two. This species is common in grassy and heathy places over much of Britain. It gets its English name from the supposed resemblance of the dark markings of the fore-wings to the face of an old crone.

The moth (*Figs 7, 8*) is on the wing in May and June and may be seen in bright sunshine, flying from flower to flower. The eggs (*Figs 1, 2*) are laid on a range of host plants, but usually either grasses or legumes such as bird's-foot trefoil (*Lotus corniculatus*), clover (*Trifolium* spp.) or medick (*Medicago* spp.). The caterpillars (*Figs 3, 4*) feed up and pupate (*Fig 6*) in a cocoon (*Fig 5*) amongst the food plant in August or September.

This species is found throughout southern and central Europe and parts of Asia.

1 Egg, natural size
2 Egg enlarged to show detail
3-4 Larvae 5 Cocoon 6 Pupa
7-8 Adult moths
Plant: Bentgrass (Agrostis sp)

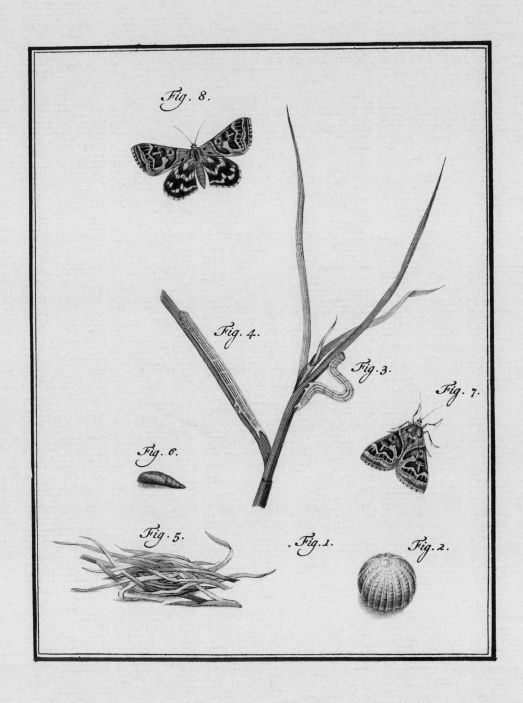

Fig. 8.

Fig. 4.

Fig. 3.

Fig. 7.

Fig. 6.

Fig. 5.

Fig. 1.

Fig. 2.

THE RED UNDERWING MOTH

NOCTUIDAE *Catocala nupta* L.

This large and handsome species is fairly common in the south of England wherever there are plenty of willow (*Salix* spp.) or poplar (*Populus* spp.) trees. The adults (*Figs 6, 7*) are on the wing in August and September and often even October. They are frequently to be found during the day sitting on fences and telegraph poles, sometimes several on one pole.

The eggs (*Figs 1, 2*) are laid on the bark of the host trees and over-winter to hatch in the following spring. The larvae (*Figs 3, 4*) are nocturnal feeders and hide in the cracks of the bark during the day. They are fully fed by July and spin a cocoon amongst the leaves in which to pupate (*Fig 5*). The moths emerge in August and September.

This species is found throughout much of Europe and temperate Asia to Japan and northern India.

1 Eggs, natural size
2 Egg enlarged to show detail 3 Larva
4 Larva 5 Pupa 6-7 Adult moths
Plant:Willow (Salix sp)

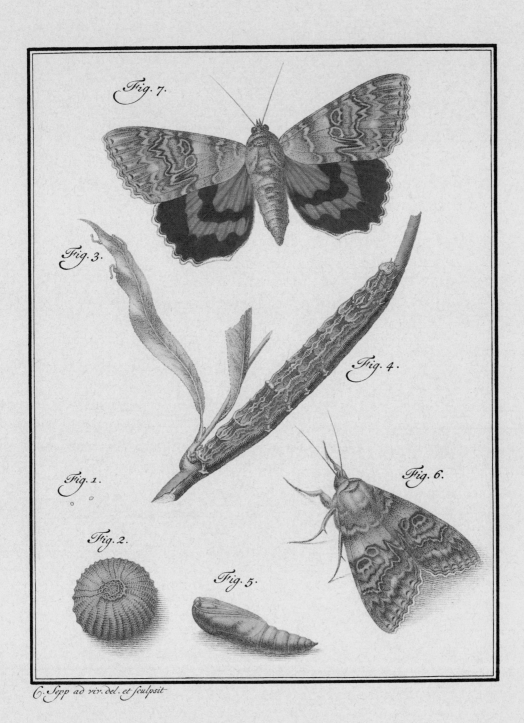

Fig. 7.

Fig. 3.

Fig. 4.

Fig. 1.

Fig. 6.

Fig. 2.

Fig. 5.

C. Sepp ad viv. del. et sculpsit

THE SNOUT MOTH

Noctuidae *Hypena proboscidalis* L.

This moth is very common wherever nettles (*Urtica dioica*) grow. The adults (*Figs 9, 10*) are active during June and July and may be flushed from nettle patches and any other rank vegetation nearby. The most obvious feature of this insect, apart from the slightly hooked delta wing shape when at rest, is the length of the palps, which gives the moth its English name. In Britain this species has normally only one generation a year, with only occasionally a small second flight in September. On the continent at the same latitude, however, it regularly has two generations.

The eggs (*Figs 1, 2*) are laid on nettles and hatch in July. The partly grown larvae overwinter (*Figs 3, 4*) and feed up in the following spring. The fully fed larvae (*Fig 6*) spin a thin cocoon (*Fig 7*) in which to pupate (*Fig 8*) in late April or early May.

This species is widely distributed in Europe and temperate Asia and is very common throughout Britain.

1 Eggs, natural size
2 Egg enlarged to show detail
3 Newly hatched larvae suspended from
threads 4-6 Larvae 7 Cocoon 8 Pupa
9-10 Adult moths
Plant:Nettle (Urtica dioica)

Fig. 10.

Fig. 4.

Fig. 5.

Fig. 6.

Fig. 9.

Fig. 3.

Fig. 7.

Fig. 8.

Fig. 2.

Fig. 1.

THE LARGE EMERALD MOTH

GEOMETRIDAE *Geometra papilionaria* L.

This species introduces the section in this book representing the large family of moths called the Geometridae, the "earth measurers". The name comes from the larval stages most of which have only the six true legs at the anterior end of the body and four claspers at the rear. This means that in order to progress they use a sort of looping movement, first gripping with the true legs, releasing the claspers, bringing the rear of the body up to the true legs and grasping with the claspers. The true legs are then released and the whole movement starts again. This tends to give the impression of measuring out the ground as the caterpillar moves over it. It also accounts for their other vernacular name of loopers.

The adults of many species in this family may be flushed out of their daytime resting places in rank vegetation or found resting on fences and tree trunks. This means that many people come across these species in a casual way much more often than they do the moths that we have been describing in the previous section of plates.

This species (*Figs 7, 8*) flies in June and July and may be found commonly in woods and on scrubby heaths and moors. The eggs (*Figs 1, 2*) are laid on the leaves of birch (*Betula* spp.) and hazel (*Corylus avellana*) and occasionally on other trees. The larvae (*Fig 3*) hatch in the autumn and hibernate as small caterpillars, clinging to a silk pad on the twigs of the host tree. At this stage they are largely brown in colour and difficult to see on the twigs, but when feeding commences again in the spring they quickly become greener (*Figs 4, 5*) so that they remain concealed as the leaves of the host tree develop. This type of colour change from brown or ochreous in the autumn to green in the spring is not uncommon in many insects. The larvae are fully fed by May and pupate (*Fig 6*) on the ground in a rudimentary cocoon between dead leaves of the food plant.

This species is found through much of northern and central Europe and temperate Asia to Japan.

1 Egg, natural size
2 Egg enlarged to show detail
3-5 Larvae 6 Pupa 7 Male moth
8 Female moth
Plant:Birch (Betula sp)

Fig. 8.

Fig. 7.

Fig. 4.

Fig. 3.

Fig. 5.

Fig. 6.

Fig. 2.

Fig. 1.

THE MALLOW MOTH

GEOMETRIDAE *Larentia clavaria* Haw.

This species and its common close relative, the Shaded Broadbar (*L. chenopodiata*) look, at first sight, like well-marked specimens of the Snout moth (*pp 196, 197*), but on closer examination it will be seen that the adults (*Figs 6, 7*) do not have the long palps. The larvae (*Figs 3, 4*) are typical loopers whereas the Snout larvae (*p 197, Fig 6*) are not.

The Mallow moth is common wherever its food plant, the common mallow (*Malva sylvestris*) grows, and will feed on garden plants such as hollyhock. The adults are on the wing in September and October and may be flushed from the food plant during the day. The eggs (*Figs 1, 2*) are laid on the food plant and overwinter to hatch in the following spring. The caterpillars (*Figs 3, 4*) feed up on the young shoots and become fully grown by late May or early June. They then pupate (*Fig 5*) and the moths emerge in September.

This species is found in areas where its food plant grows over much of central and southern Europe and parts of temperate Asia.

1 Egg, natural size
2 Egg enlarged to show detail
3-4 Larvae 5 Pupa 6-7 Adult moths
Plant:Mallow (Malva sylvestris)

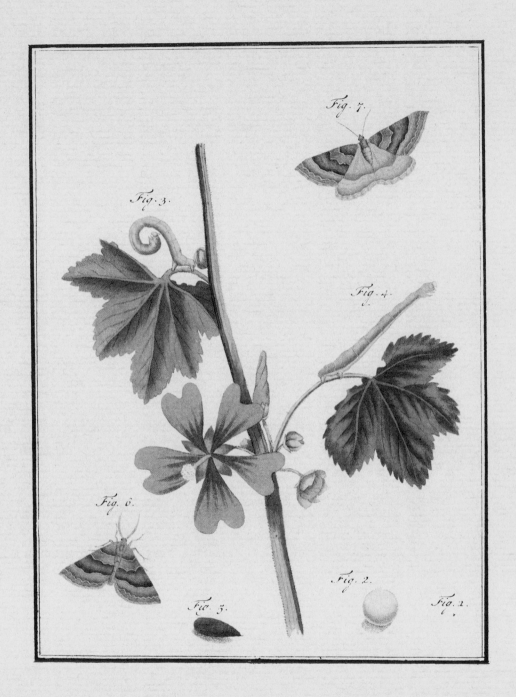

Fig. 7.

Fig. 3.

Fig. 4.

Fig. 6.

Fig. 5.

Fig. 2.

Fig. 1.

THE STREAK MOTH

GEOMETRIDAE *Chesias legellata* Schiff.

This species is common throughout much of Britain wherever broom (*Sarothamnus scoparius*) grows. There is also a similar but smaller species, the Broom-tip (*C. rufata*), which is much more local and southern in its distribution.

The Streak (*Figs 3, 4*) is on the wing in the autumn from September to early November. The eggs are laid on the young shoots of broom and they overwinter there. The larvae (*Fig 1*) hatch in April and feed on the leaves and flowers. These larvae are fully fed by late May or early June, when they drop to the ground and pupate (*Fig 2*) in the soil. The adults emerge in the autumn. The parasite pupae shown in *Figure 5* are probably those of a species of braconid wasp in the genus *Apanteles* (a very common type of parasite in Lepidoptera) or else in the genus *Microplitis*. The puparium in *Figure 6* is probably that of a parasitic fly in the family Tachinidae.

The Broom-tip has a more complex life cycle in that adults are on the wing in June, July and August, and small numbers are also found in the spring from March to June. In this species the larvae feed on broom in July, August and September and the pupae overwinter. Most emerge the following June and July, but a small percentage remains in the pupa for a second winter to emerge in the earlier part of the year.

Both species are found where their food plant grows over much of southern and central Europe and North Africa.

1 Larva 2 Pupa 3-4 Adult moths
5 Pupae of parasitic braconid wasp
6 Puparium of parasitic fly
Plant:Broom (Sarothamnus scoparius)

202

Fig. 4.

Fig. 1.

Fig. 3.

Fig. 6.

Fig. 2.

Fig. 5.

THE WINTER MOTH

GEOMETRIDAE *Operophthera brumata* L.

This species is probably one of the most studied of our moths, largely from research carried out at Wytham Woods near Oxford beginning in the late 1940s. The adults (*Figs 7, 8, 9*) are active in the late autumn and early winter from late October until the end of December. The males (*Figs 7, 8*) fly at dusk and may be seen in large numbers in woods and orchards throughout the country. The females (*Fig 9*) are completely flightless, as is common in many species of moths active during the winter and early spring. The females walk up the trunks of trees during the night and the males mate with them on the lower parts of the trunks. The male sits head downward behind the female during mating. The females continue up the trees to lay their eggs (*Figs 1, 2*) in cracks and crevices in the upper parts of the tree.

These eggs hatch in April and the larvae (*Figs 3, 4*) feed on the leaves of a wide variety of trees, e.g. oak (*Quercus* spp.), apple (*Malus sylvestris*) birch (*Betula* spp.), etc., and may completely defoliate individual or small groups of trees. To prevent this, apple trees in orchards are often grease banded in winter to catch the flightless females on their way up the trunks. The caterpillars are fully fed by late May and drop to the ground to pupate (*Fig 6*) in a cocoon (*Fig 5*) in the soil.

This species has a very wide distribution covering northern and central Europe, temperate Asia, and North America. Throughout its range it is a pest of orchard crops. Several parasites have been introduced into North America from Europe, with some success, for the biological control of this pest.

1 Eggs, natural size
2 Egg enlarged to show detail
3-4 Larvae 5 Cocoon 6 Pupa
7-8 Male moths
9 Flightless female moths
Plant:Apple (Malus sp)

THE PHOENIX MOTH

GEOMETRIDAE *Lygris prunata* L.

This species, with its beautiful shades of brown in the wings, is a familiar insect to those with fruit gardens. It is very common in gardens and hedges where any type of gooseberries or currants (*Ribes* spp.) grow.

In July and August the adults (*Figs 6, 7*) are on the wing throughout the British Isles. The eggs (*Figs 1, 2*) are laid on the twigs of members of the genus *Ribes*. The stick-like caterpillars (*Figs 3, 4*) hatch in the following spring and are usually fully fed by mid-June. They pupate (*Fig 5*) between two leaves on the food plant. The leaves are held together with a few strands of silk.

The moth may easily be netted as it flies along hedges visiting flowers at dusk. It is common throughout Britain and has a circumpolar distribution, being found in northern and central Europe and Asia and in the northern USA and Canada.

1 Egg, natural size laid on twig
2 Egg enlarged to show detail
3-4 Larvae 5 Pupa 6-7 Adult moths
Plant:Currant (Ribes sp)

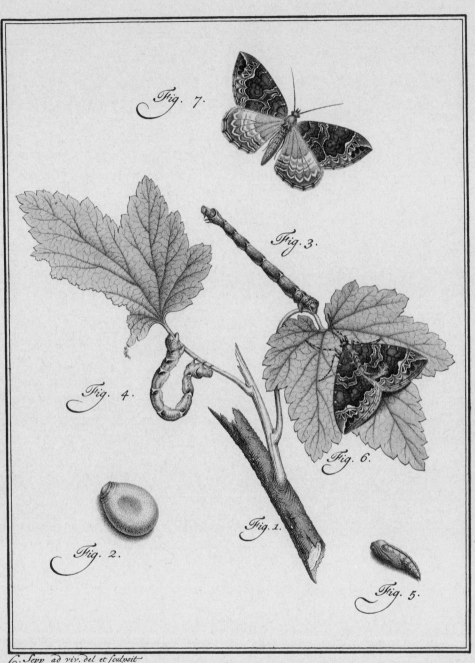

Fig. 7.

Fig. 3.

Fig. 4.

Fig. 6.

Fig. 1.

Fig. 2.

Fig. 5.

C. Sepp ad viv. del et sculpsit

THE SPINACH MOTH

GEOMETRIDAE *Eulithis mellinata* Fab.

The illustration (*Fig 6*) of the adult of this common garden-living moth is much more heavily marked than most British specimens, which tend to have the markings on the fore-wings reduced to the three darkest cross-bands. The English name of this species comes about because during the day, it is often found sitting on spinach, a vegetable frequently grown near to the currant bushes on which the larvae feed.

The eggs (*Figs 1, 2*) are of similar flask-like shape to those of the Phoenix (*pp 206, 207*) and are laid on the tips of the twigs of both red and black currants (*Ribes rubrum, R. nigrum*). The eggs overwinter and hatch in the following April or May.

The caterpillars (*Fig 3*) feed at night on the leaves of the food plant and pupate (*Figs 4, 5*) in June. There is only one generation per year, the adults of which (*Figs 6, 7*) are on the wing in July and August. They are easily attracted to light and being primarily a garden species may often be found fluttering round street lights.

This species is common throughout England and southern Scotland and is found across western and central Europe eastwards to Mongolia.

1 Egg, natural size
2 Egg enlarged to show detail 3 Larva
4-5 Pupae 6-7 Adult moths
Plant:Currant (Ribes sp)

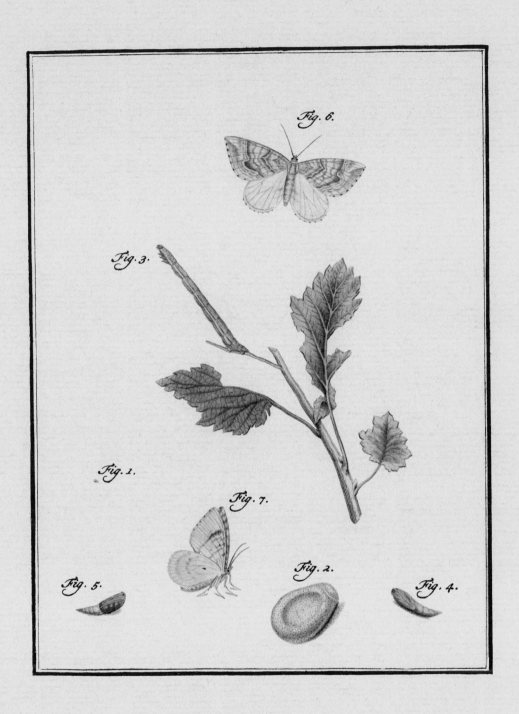

Fig. 6.

Fig. 3.

Fig. 1.

Fig. 7.

Fig. 5.

Fig. 2.

Fig. 4.

THE COMMON MARBLED CARPET MOTH

GEOMETRIDAE *Chloroclysta truncata* Hufn.

This species is one of three closely related species found in the British Isles, the Common Marbled Carpet (*C. truncata*), the Dark Marbled Carpet (*C. citrata*) and the Northern Marbled Carpet (*C. coccinata*). The first two species are very common throughout Britain, but the third is confined to high ground in the highlands and islands of western Scotland.

The two common species are both very variable in colour and pattern as adults. Some idea of the range in the Common Marbled Carpet may be gained from the forms illustrated in *Figures 9, 10* and *11*. The Dark Marbled Carpet is very similar and has a similar range of colour forms; the genitalia will reliably tell the adults apart. The Dark Marbled Carpet, however, usually has paler hind-wings and the discal spot is less well marked than in the Common Marbled Carpet. The most reliable character, however, is in the fore-wings, on which, in the Dark Marbled Carpet, the outline of the outer boundary of the main cross-bar has a sharper and larger projection just in front of its mid-point.

The flight times of the adults are different in the two species: the Common Marbled Carpet is double-brooded with the adults on the wing in May and June and again in August and September; the Dark Marbled Carpet is single-brooded and flies in July and August.

The eggs are laid on the leaves of a variety of plants, e.g., sallow (*Salix* spp.), birch (*Betula* spp.) and blaeberry (*Vaccinium myrtillis*), but the species is most often associated with strawberry (*Fragaria vesca*). The caterpillars (*Fig 7*) of the first generation feed up rapidly and spin one or two leaves together as a shelter and pupate (*Fig 8*) in July. Those of the second generation hatch in the autumn and hibernate as small larvae, which feed up in the spring and pupate in May.

The Dark Marbled Carpet feeds on much the same range of plants, but the eggs in this species overwinter. The caterpillars are similar to those of the Common Marbled Carpet but without the red dashes.

Both these species are common in gardens and woodland edges and are found abroad in northern and central Europe and Asia. The Dark Marbled Carpet is also found in North America.

7 Larva 8 Pupa 9-11 Adult moths
Plant:Wild strawberry (Fragaria vesca)

Fig. 9.

Fig. 10.

Fig. 7.

Fig. 11.

Fig. 8.

THE SILVER GROUND CARPET MOTH

GEOMETRIDAE *Xanthorhoë montanata* Bkh.

This moth is very common in gardens, hedgerows, woodland edges and waste ground. It is one of a large group of species with similar basic wing patterns known as the "carpet" moths. Several of this group are very common, but this species and its smaller and darker relative, the Garden Carpet (*X. fluctuata*), are perhaps the most abundant and familiar.

In June and July the adults (*Figs 11, 12*) of the Silver Ground Carpet may be flushed from their daytime resting places in dense vegetation such as nettle beds or the herbaceous border. The intensity of the pattern on the fore- and hind-wings is variable and many specimens are not as heavily marked as those illustrated. The central band, however, is always well marked and may be any colour from a light sandy shade to a dark grey-brown.

The eggs (*Figs 1, 2, 3*) are laid on a variety of grasses and broad-leaved plants, often on primroses and polyanthus (*Primula* spp.). The caterpillars (*Fig 4*) hatch and overwinter as small larvae. They feed up quickly (*Figs 5, 6, 7*) in the early spring and pupate (*Fig 9*) in late April or early May. The pupae are found in a slight cocoon inside a curled leaf.

This species is distributed throughout temperate Europe and Asia as far as western Siberia.

1 Eggs, natural size
2-3 Eggs enlarged to show detail
4-7 Larvae 8 Cocoon 9 Pupa
10 Cremaster (Pupal hook-like
attachment)
11-12 Adult moths
Plant:Polyanthus (Primula sp)

Fig. 12.

Fig. 6.

Fig. 7.

Fig. 5.

Fig. 11.

Fig. 4.

Fig. 10.

Fig. 9.

Fig. 8.

Fig. 1.

Fig. 2.

Fig. 3.

THE MAGPIE MOTH

GEOMETRIDAE *Abraxas grossulariata* L.

This pretty species is a common pest in the fruit garden, where it feeds on gooseberry and currants, which it may completely defoliate. It has a wide range of other food plants and in the north of Scotland it normally feeds on heather (*Calluna vulgaris*), though it still prefers *Ribes* spp. when they are available.

There is one generation of this insect in the year, the adults (*Figs 5, 6*) being on the wing in July and August. The amount of black on the wings is very variable and in some specimens may cover most of the surface of the wings. The ground colour of the wings may also vary between white and rich cream.

The eggs (*Figs 1, 2*) are laid in small groups on the leaves of the food plant. The young caterpillars hatch in August and overwinter as young larvae. They begin feeding again in the early spring and are fully grown (*Fig 3*) by late May. They then spin a light web under a leaf and pupate (*Fig 4*).

This species is common in gardens, hedgerows, and along moorland edges throughout Britain. Abroad, it is found across most of temperate Europe and Asia.

1 Eggs, natural size
2 Egg enlarged to show detail 3 Larva
4 Pupa 5-6 Adult moths
Plant:Currant (Ribes sp)

Fig. 6.

Fig. 3.

Fig. 1.

Fig. 2.

Fig. 5.

Fig. 4.

THE LARGE THORN MOTH

GEOMETRIDAE *Ennomos autumnaria* Wernb.

This insect is a representative of a group of species with wavy-edged wings called thorn moths because of their stick-like caterpillars. The commonest species of this group in Britain is the Canary-shouldered Thorn (*E. alnaria*) which is like a smaller version of the Large Thorn but with a pretty bright yellow area of 'fur' on the head and thorax.

The Large Thorn is a fairly rare visitor to the southern coastal countries although it is common on the other side of the Channel and right across Europe and temperate Asia and also in North America. The adults (*Figs 5, 6*) are on the wing in August and September and lay their eggs on a variety of trees. The eggs (*Figs 1, 2*) overwinter and hatch in the following spring. The caterpillars (*Fig 3*) feed up during the summer and pupate (*Fig 4*) in July and August.

The Canary-shouldered Thorn has a very similar life cycle and feeds on much the same range of host plants. It is common in damp woodlands over much of Britain.

1 *Eggs, natural size*
2 *Egg enlarged to show detail 3 Larva*
4 *Pupa 5 Female moth 6 Male moth*
Plant:Elm (Ulmus sp)

216

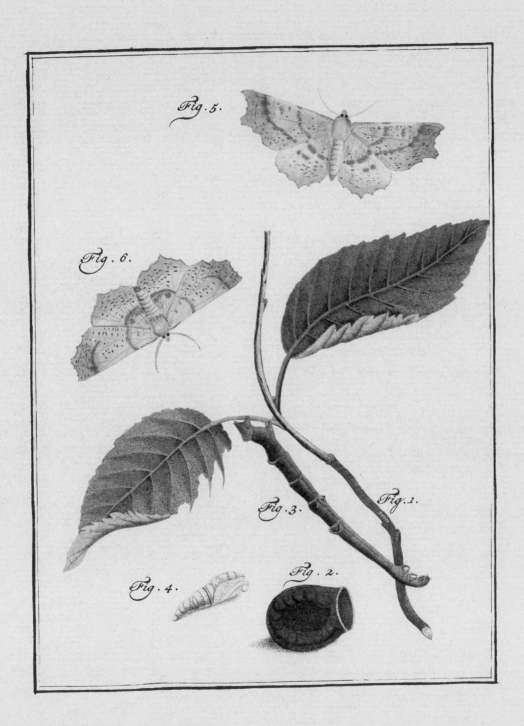

Fig. 5.

Fig. 6.

Fig. 3.

Fig. 1.

Fig. 4.

Fig. 2.

THE SWALLOWTAILED MOTH

GEOMETRIDAE *Ourapteryx sambucaria* L.

This large pale-coloured species is common in gardens, woodland edges and scrubland. It is very conspicuous as it flies slowly in the dusk. The adults (*Figs 11, 12*), unlike any other common species of moth in this country, have short tails to the hind-wings. There is one generation in the year, which is on the wing in July.

The eggs (*Figs 1, 2, 3, 4*) are laid in small batches on elder (*Sambucus nigra*), hawthorn (*Crataegus oxyacanthoides*), ivy (*Hedera helix*) and other plants. The eggs hatch and the caterpillars feed up in the autumn (*Figs 5, 6*), overwinter and complete their development (*Figs 7, 8*) in the spring. They are fully fed by May and the caterpillars spin a hammock-like cocoon (*Fig 9*) on the food plant in which to pupate (*Fig 10*).

The adults emerge from these pupae after six or seven weeks. In some years, a few caterpillars will feed up, pupate and emerge in the late autumn (September and October), but this is a rare occurrence in this country.

The moth is common in England and southern Scotland and is found abroad through much of southern and central Europe and temperate Asia to Japan.

1-3 Eggs 4 Egg enlarged to show detail
5-8 Larvae 9 Cocoon 10 Pupa
11-12 Adult moths
Plant:Elder (Sambucus nigra)

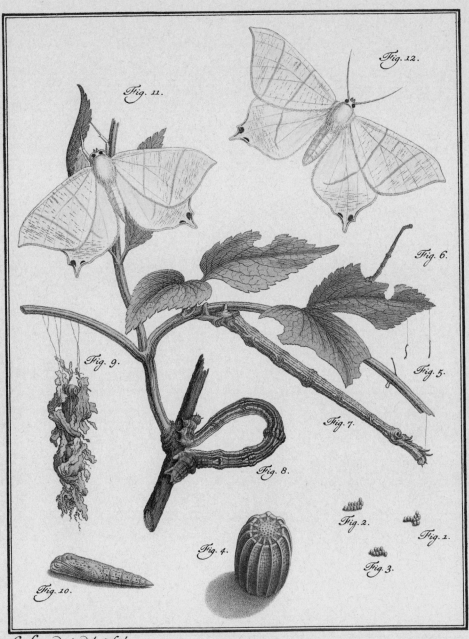

Fig. 12.

Fig. 11.

Fig. 6.

Fig. 9.

Fig. 5.

Fig. 7.

Fig. 8.

Fig. 2.

Fig. 1.

Fig. 4.

Fig. 3.

Fig. 10.

C. Sepp ad viv. del. et sculps.

THE BRIMSTONE MOTH

GEOMETRIDAE *Opisthograptis luteolata* L.

This pretty and distinctive species is a common inhabitant of gardens and woodland edges over the whole of Britain. The adults (*Figs 7, 8*) may be found on the wing at any time from May to August, though the greatest number fly in May and June. There are at least two generations in the year, but they overlap very widely.

The eggs (*Figs 1, 2*) are laid in small groups on a variety of types of bushes, but this species is most often associated with hawthorn (*Crataegus oxyacanthoides*). The caterpillars (*Figs 3, 4*) hatch within a short time and some of them feed up quickly and pupate in late June or early July to emerge as adults a few weeks later. Others, even in the same egg batch, feed slowly and overwinter as larvae before pupating (*Fig 6*) in a cocoon near the ground. The young of the July and August adults either overwinter as larvae (the slow-growing morph) or as pupae (the fast-growing morph).

The markings on the head of the caterpillar (*Fig 5*) are interesting in that they give it the look of a small mammal.

This species is found commonly throughout south and central Europe, temperate Asia and North Africa.

1 Eggs, natural size
2 Egg enlarged to show detail
3-4 Larvae
5 Detail of head-markings of larva
6 Pupa 7-8 Adult moths
9 Side view of head of larva, showing eyes
Plant:Hawthorn (Crataegus sp)

Fig. 8.

Fig. 4.

Fig. 3.

Fig. 7.

Fig. 1.

Fig. 5.

Fig. 9.

Fig. 2.

Fig. 6.

THE BRINDLED BEAUTY MOTH

GEOMETRIDAE *Biston hirtarius* Clerk.

This early spring species is common throughout Britain and is even present in the parks of central London. In March and April the adults (*Figs 9, 10, 11*) may be found sitting on the bark of trees or around lampposts. The males (*Fig 11*) have large feathery antennae, while the females (*Figs 9, 10*) have filiform antennae. The colour of the adults varies from fairly light, as in *Figure 10*, to almost black, the dark individuals being much commoner in towns.

The eggs are laid in small groups on the bark of a range of trees, e.g., lime (*Tilia* spp.), elm (*Ulmus* spp.) or plum (*Prunus* spp.). The eggs hatch in late May and the caterpillars (*Fig 7*) feed until July on the leaves of the trees. When fully grown, they descend from the trees and pupate (*Fig 8*) in the soil at the base.

This species is found across Europe and temperate Asia.

7 Larva 8 Pupa 9 Female moth
10 Light-coloured female moth
11 Male moth
Plant:Plum (Prunus sp)

Fig. 10.

Fig. 11.

Fig. 7.

Fig. 9.

Fig. 8.

THE OAK BEAUTY MOTH

GEOMETRIDAE *Biston stratarius* Hufn.

This is another early species common in woodlands through much of England, and again it may be seen by day sitting on the trunks of the trees. The pretty adults (*Figs 4, 5*) are on the wing in March and April. The size of the black bands and the degree of overall speckling of the wings varies from specimen to specimen.

The eggs are laid on the bark of oak (*Quercus* spp.) or birch (*Betula* spp.). The caterpillars (*Figs 1, 2*) hatch in early May and, after feeding up, descend to pupate in late June or early July. The pupae (*Fig 3*) are found underground near the host tree and overwinter to emerge as adults in the following spring.

This species is locally common through most of England and is found throughout south and central Europe and Asia Minor.

1-2 Larvae 3 Pupa 4 Male moth
5 Female moth
Plant:Oak (Quercus sp)

Fig. 5.

Fig. 4.

Fig. 1.

Fig. 2.

Fig. 3.

THE PEPPERED MOTH

GEOMETRIDAE *Biston betularius* L.

This common and well-known species has been much studied because it provides one of the most spectacular and well-documented cases of industrial melanism. The typical form of the adults, after which the moth gets its vernacular name, is illustrated in *Figures 9, 10* and *11*. The completely black form (*carbonaria*) was quite unknown in Sepp's day and was first noted in Manchester in 1850. It quickly spread through much of industrial Britain and black moths also began to be noted in some areas on the continent, e.g., in Holland in 1890.

This black form is now the most common form over much of the country, even occurring in many rural areas, as an appreciable proportion of the population. It is genetically dominant to the light form and has been selected out by differential predation. The light form is well camouflaged when at rest on the trunk of a lichened tree in the country, but when sitting on a lichen free and smoke blackened tree in the city, it is very obvious to bird predators. The black forms, on the other hand, are well camouflaged in the city. This tips the balance of natural selection towards the black form in the polluted areas, though the light form will still be at an advantage in the country areas.

The adult moths are on the wing in May and June and may be found sitting on tree trunks and fences during the day. The eggs (*Figs 1, 2*) are laid on a variety of trees. The caterpillars (*Figs 3, 4, 5, 6, 7*) feed up rapidly, becoming fully fed by September. Viewed from the front the shape of the head (*Figs a, b*) is very cat-like, the two projections looking like ears.

The pupa (*Fig 8*) is found in the soil at the base of the host tree and overwinters to emerge as the adult in the following summer.

This species is common in much of Europe and Asia as far east as China and Japan.

1 Eggs, natural size
2 Egg· enlarged to show detail
3-6 Larvae a-b Details of head
Plant:Alder (Alnus glutinosa)

Fig. 3.

Fig. 4.

Fig. 6.

Fig. 5.

Fig. 1.

Fig. 2.

b

a

THE PEPPERED MOTH

GEOMETRIDAE *Biston betularius* L.

(continued)

7 Larva 8 Pupa 9 Male moth
10 Female moth 11 Female moth
Plant: Alder (Alnus glutinosa)

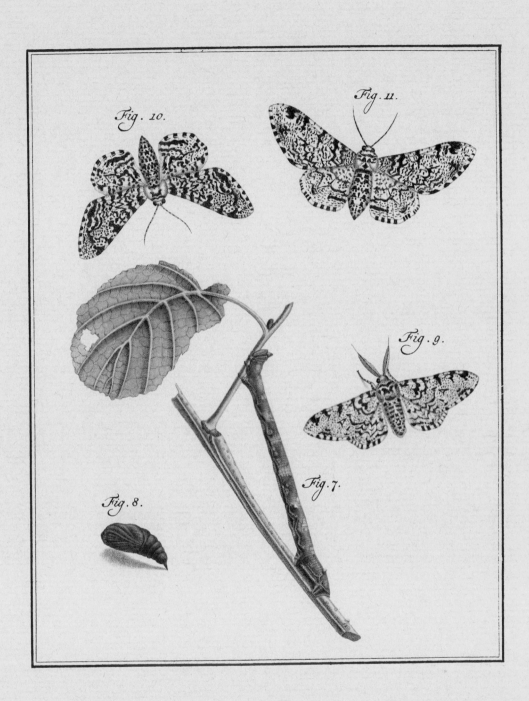

Fig. 10.

Fig. 11.

Fig. 9.

Fig. 7.

Fig. 8.

THE BORDERED WHITE, OR PINE LOOPER MOTH

GEOMETRIDAE *Bupalus piniaria* L.

This pretty species is very common in pine woods, and occasionally also on larch. It is sometimes so numerous as to defoliate the trees and, because it is a pest in many parts of Europe, it has been the subject of a number of intensive studies over many years, especially in Holland and Germany.

The adult moths (*Figs 8, 9, 10*) are largely day-flying and emerge from the pupae in May and June, usually in the early morning. The males (*Figs 8, 9*) are very different in colour from the females (*Fig 10*) and one finds a gradation in the colour of both sexes as one moves from south to north through Britain. The ground colour of the males changes from yellow to white while the females change from a generally orange colour to a much darker brown.

The eggs (*Figs 1, 2*) are laid in small linear clusters of up to 25 on the needles of the host tree, usually scots pine (*Pinus sylvestris*). They hatch in about 20 days and the small caterpillars (*Figs 3, 4*) begin to eat small nicks and grooves in the needles. They feed until October or November, by which time they are fully fed (*Figs 5, 6*). They then fall from the tree and pupate in the litter layer in a small silk-lined chamber. The pupae (*Fig 7*) overwinter and emerge as adults the next summer.

This species is common in pine woods, natural and man-made, throughout Britain, and is found in northern and central Europe and temperate Asia to eastern Siberia.

1 Eggs, natural size
2 Egg enlarged to show detail
3-6 Larvae 7 Pupa 8 Male moth
9 Male moth 10 Female moth
Plant:Pine (Pinus sylvestris)

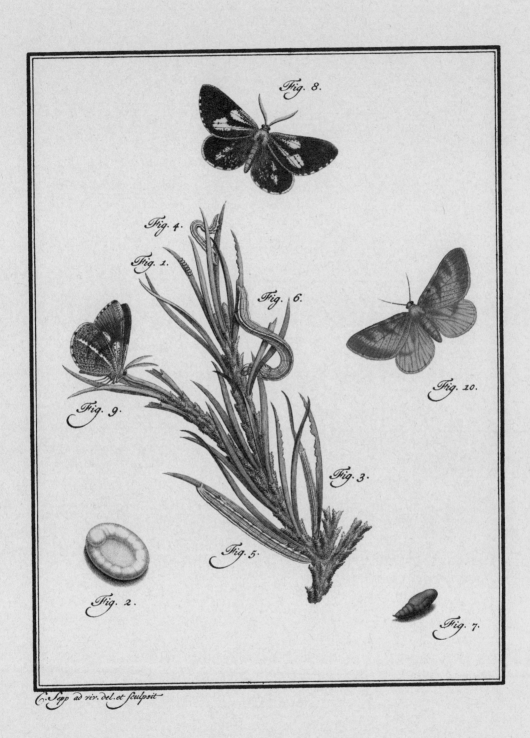

Fig. 8.

Fig. 4.

Fig. 1.

Fig. 6.

Fig. 9.

Fig. 10.

Fig. 3.

Fig. 2.

Fig. 5.

Fig. 7.

C. Sepp ad viv. del. et sculpsit.

THE SIX-SPOT BURNET MOTH

ZYGAENIDAE *Zygaena filipendulae* L.

This day-flying moth must be familiar to everyone who has taken a holiday by the sea, as it is very common in sand dunes and on cliff tops. It is by far the commonest of our green and red burnet moths, most of the others being very local in their occurrence. Three species have semi-transparent wings, but the general pattern of spots is similar in all seven species. The species with translucent wings are: the Transparent Burnet (*Z. purpuralis*) found in Cornwall, Wales and west Scotland, the Mountain Burnet (*Z. exulans*) found in the Cairngorms, and the New Forest Burnet (*Z. meliloti*) found in the New Forest as its name suggests. Two species lack one of the outer pair of red spots on the fore-wings: the Five-spot Burnet (*Z. trifolii*) found in damp meadows in southern England and the Narrow-bordered Five-spot Burnet (*Z. lonicerae*) found in similar situations but extending as far north as the Lake District. The final two species, the Six-spot Burnet (*Z. filipendulae*) and the Northern Burnet (*Z. achilleae*), are very much alike; the latter is more hairy in the body and is confined to parts of the Western Isles and western Highlands of Scotland.

The Six-spot Burnet adults (*Figs 8, 9*) are on the wing in grassy places throughout Britain in July and August. The eggs (*Figs 1, 2*) are laid on various legumes, but most often on bird's-foot trefoil (*Lotus corniculatus*) and kidney vetch (*Anthylis vulneraria*). The caterpillars (*Figs 3, 4*) feed up slowly and overwinter as small larvae to complete their growth in the following spring. They are fully grown by late May and spin their papery yellow cocoons (*Fig 5*) on any tall vegetation. These cocoons and the pupae (*Fig 6*) they contain are often torn open by birds if a convenient perch is available. The moths emerge in July by tearing open the front end of the cocoon (*Fig 7*).

1 Eggs, natural size
2 Egg enlarged to show detail
3-4 Larvae 5 Cocoon 6 Pupa
7 Cocoon after emergence of adult
8-9 Adult moths
Plant:Bird's foot trefoil (Lotus corniculatus)

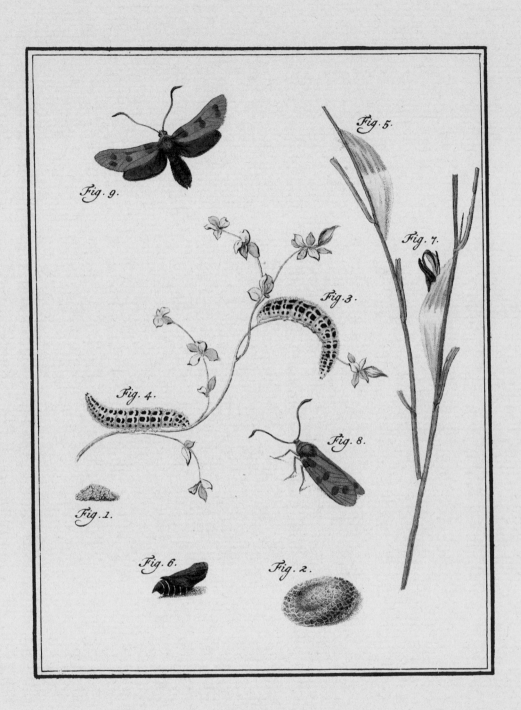

Fig. 9.

Fig. 5.

Fig. 7.

Fig. 3.

Fig. 4.

Fig. 8.

Fig. 1.

Fig. 6.

Fig. 2.

THE HORNET CLEARWING MOTH

Sesidae *Sesia apiformis* Cl.

Most members of this family of moths look like bees or wasps and the present species is a good mimic of the hornet. It is a day-flyer with a rapid flight and hence is difficult to follow when in flight.

The adults of this species (*Figs 6, 7*) are on the wing in hot, sunny weather in June and July. The eggs are laid on the bark at the base of poplar trees (*Populus* spp.). The very similar Lunar Hornet Clearwing (*S. crabroniformis*) lives in sallow (*Salix* spp.). The larvae live in the wood for two years (*Figs 1, 2, 3*), finally pupating (*Fig 4*) in May of the second year in a cocoon of wood chippings. The moth emerges, usually in the afternoon, leaving the pupal case (*Fig 5*) protruding from the hole in the trunk of the tree.

This species is local and most abundant in eastern England, just reaching the Scottish border. The Lunar Hornet Clearwing is rather more widely distributed in wet habitats throughout England and southern Scotland. Abroad the Hornet Clearwing is found through much of Europe and in North America, while the Lunar Hornet Clearwing is much more restricted to central and western Europe.

1 Larva in burrow 2-3 Larvae 4 Pupa
5 Empty pupal case 6 Male moth
7 Female moth
Plant:Trunk of Poplar tree (Populus sp)

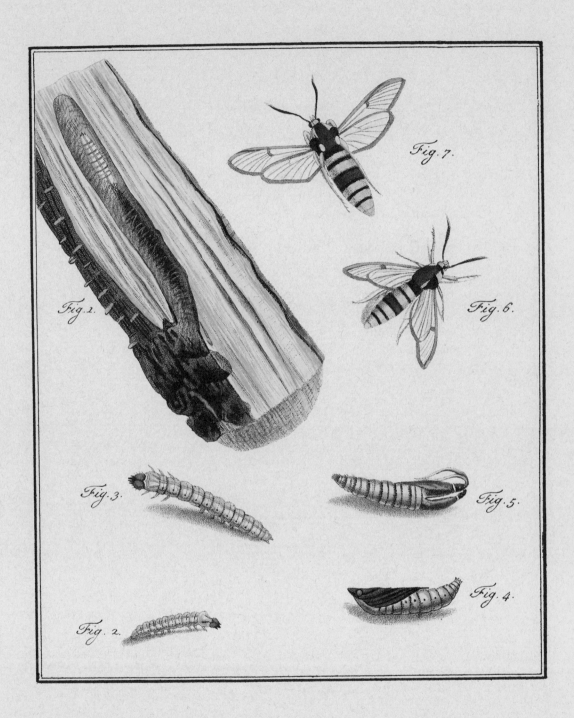

Fig. 7.

Fig. 6.

Fig. 1.

Fig. 3.

Fig. 5.

Fig. 4.

Fig. 2.

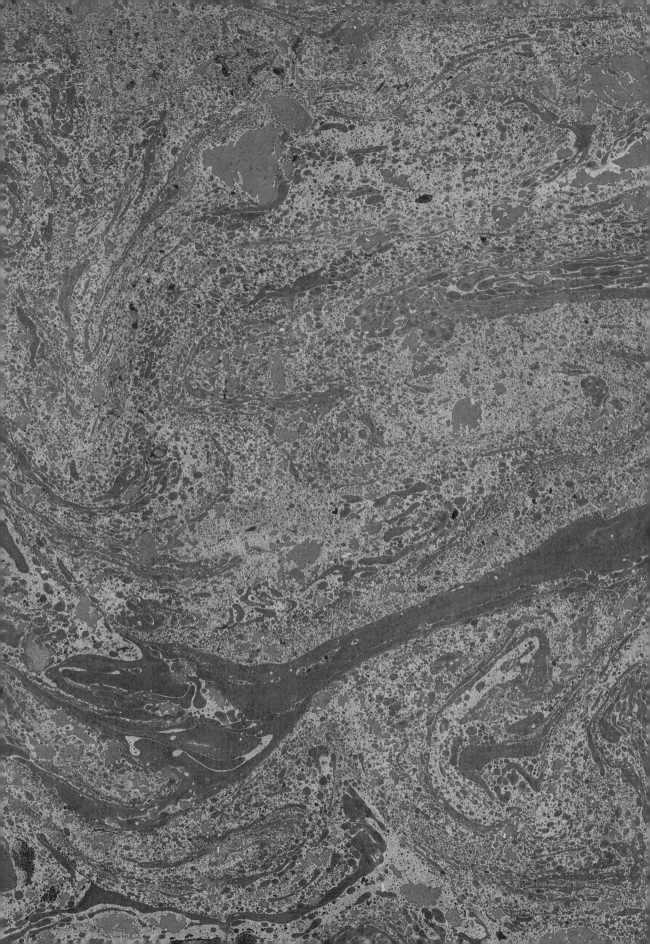